CW01509694

CHINA'S RISING SEA POWER

Since the collapse of Soviet naval power, China has become the operator of the world's largest tactical submarine fleet. As the world wonders what Beijing intends to do with its increasingly powerful and effective undersea warfare capacity, what does this mean for strategic stability in East Asia? This book addresses these questions by exploring similarities between China's strategic outlook today and that of earlier continental powers whose submarine fleets challenged dominant maritime powers for regional hegemony. Using insights from classical naval strategic theory it examines Beijing's strategic logic in making tactical submarines the keystone of China's naval force structure, and investigates the influence of Soviet naval strategy and ancient Chinese military thought on the PLA Navy's strategic culture. It finally contends that China's increasingly capable submarine fleet could play a key role in Beijing's use of force to resolve the Taiwan issue. With attention focused on China's missile build-up opposite Taiwan, this timely volume reminds us that there is real danger in underestimating the potential of the PLA Navy's submarines to destabilise any future Taiwan Strait crisis.

This book will be of foremost interest to students of Chinese politics, Asian Security Studies and international relations in general, as well as to professionals in these fields.

Peter Howarth is a diplomat with the Australian Department of Foreign Affairs and Trade, where he was Director of Nuclear Non-proliferation Policy. Most recently he served as a senior strategic analyst with Australia's foreign intelligence assessment agency. He holds a PhD from the University of Paris.

ASIAN SECURITY STUDIES
Series Editors: Sumit Ganguly, Indiana University, Bloomington
and Andrew Scobell, US Army War College

Few regions of the world are fraught with as many security questions as
Asia. Within this region it is possible to study great power rivalries, irreden-
tist conflicts, nuclear and ballistic missile proliferation, secessionist
movements, ethnoreligious conflicts and inter-state wars. This new book
series will publish the best possible scholarship on the security issues
affecting the region, and will include detailed empirical studies, theoretically
oriented case studies and policy-relevant analyses as well as more general
works.

CHINA AND INTERNATIONAL INSTITUTIONS
Alternate Paths to Global Power
Marc Lanteigne

CHINA'S RISING SEA POWER
The PLA Navy's Submarine Challenge
Peter Howarth

IF CHINA ATTACKS TAIWAN
Military Strategy, Politics and Economics
Steve Tsang

CHINESE CIVIL-MILITARY RELATIONS
The Transformation of the People's Liberation Army
Nan Li

THE CHINESE ARMY TODAY
Tradition and Transformation for the 21st Century
Dennis J. Blasko

TAIWAN'S SECURITY
History and Prospects
Bernard D. Cole

CHINA'S RISING SEA POWER

The PLA Navy's Submarine Challenge

Peter Howarth

Routledge
Taylor & Francis Group

LONDON AND NEW YORK

First published 2006
by Routledge
2 Park Square, Milton Park, Abingdon, Oxon, OX14 4RN

Simultaneously published in the USA and Canada
by Routledge
270 Madison Ave, New York, NY 10016

Routledge is an imprint of the Taylor & Francis Group

Transferred to Digital Printing 2009

© 2006 Peter Howarth

Typeset in Times New Roman by Taylor & Francis Books

British Library Cataloguing in Publication Data
A catalogue record for this book is available from the British Library

Library of Congress Cataloging in Publication Data
A catalog record for this book has been requested

ISBN10: 0-415-36891-X (hbk)
ISBN10: 0-415-49516-4 (pbk)

ISBN13: 978-0-415-36891-9 (hbk)
ISBN13: 978-0-415-49516-5 (pbk)

FOR ROBYN, FREYA AND CLAUDIA

CONTENTS

ACKNOWLEDGEMENTS

I would like to thank Mme Marie-Françoise Mondeil of the Ecole des Hautes Etudes Internationales (EHEI) in Paris for her encouraging comments on the original French manuscript. I would also like to thank the library staff in the Centre de documentation of the Institut des Hautes Etudes de Défense Nationale (IHEDN), which incidentally owes its existence to Admiral Castex, for their help in locating sources for this study. Above all, I would like to thank my wife, Robyn, for her support during the 'sabbatical' afforded me by her posting to Paris. Without her support and encouragement this work could not have been undertaken, let alone completed.

The opinions expressed in this book are wholly those of the author: they should not therefore be imputed to any person or organisation with whom he is associated.

LIST OF ABBREVIATIONS

ADS	Advanced Deployable System
AEW	Airborne Early Warning
AMS	Academy of Military Sciences
ASEAN	Association of Southeast Asian Nations
ASW	Anti-Submarine Warfare
AWAC	Airborne Warning and Control System
BNS	Beidou Navigation System
C^{4}ISR	Command, Control, Communications, Computers, Intelligence, Surveillance and Reconnaissance
CAPTOR	En*cap*sulated *Tor*pedo
CASC	China Aerospace S&T Corporation
CCP	Chinese Communist Party
CMC	Central Military Commission
DARPA	Defence Advanced Research Projects Agency
ELINT	Electronic Intelligence
EO	Electro-optical
EORSAT	ELINT Ocean Reconnaissance Satellite
ESM	Electronic Surveillance Measures
GLONASS	Global Navigation Satellite System
GMTI	Ground Moving Target Indication
GPS	Global Positioning System
GSD	General Staff Department
ICBM	Intercontinental Ballistic Missile
INS	Inertial Navigation System
ISR	Intelligence, Surveillance and Reconnaissance
IUSS	Integrated Undersea Surveillance System
MMA	Multi-mission Maritime Aircraft
NRO	National Reconnaissance Office
OTHR	Over-the-Horizon Radar
PLA	People's Liberation Army
PLAAF	People's Liberation Army Air Force
PRC	People's Republic of China

ROCN	Republic of China Navy
RORSAT	Radar Ocean Reconnaissance Satellite
SAM	Surface-to-Air Missile
SAR	Synthetic Aperture Radar
SBR	Space-Based Radar
SIGINT	Signals Intelligence
SLBM	Submarine-Launched Ballistic Missile
SLOC	Sea Line of Communication
SONAR	Sound Navigation and Ranging
SOSUS	Sound Surveillance System
SSB	Ballistic Missile Submarine
SSBN	Nuclear Ballistic Missile Submarine
SSGN	SSN with dedicated non-ballistic missile launchers
SSK	Hunter Killer Submarine
SSN	Nuclear Submarine
TEL	Transporter-Erector-Launcher
UAV	Unmanned Air Vehicle

INTRODUCTION

Since the end of the Cold War the focus of concerns about maritime security has shifted from the threat of disruption to seaborne trade resulting from armed interstate conflict at sea towards threats posed by non-state actors in the form of piracy, illegal fishing, smuggling of people, drugs and arms and, particularly since September 2001, terrorism.

There is good reason for this concern. The incidence of piracy, notably in Southeast Asian waters, has increased significantly, particularly since the 1997–8 Asian financial crisis. The International Maritime Bureau's annual piracy report for 2002 recorded a worldwide total of 370 attacks in 2002 compared with 335 in 2001. The highest number of attacks occurred in Indonesian waters (International Chamber of Commerce 2003). Terrorists have demonstrated dramatically the feasibility of attacking ships in ports or at sea with the suicide boat attacks against the USS *Cole* in the port of Aden in October 2000 and the attack against the French-registered tanker *Limburg* in the Gulf of Aden two years later. It would not be beyond the means of a well-organised terrorist group to use the tools and methods of Southeast Asia's pirates to attack shipping in the narrow waters of the Malacca, Sunda or Lombok Straits and cause major disruption to the maritime commerce on which the economies of Northeast Asia are so heavily dependent. Even more worrying is the prospect of terrorists using a ship as a delivery vehicle for weapons of mass destruction, detonating such a device in one of the major port cities in Asia such as Singapore, Hong Kong or Tokyo. Attacks of this kind could seriously undermine the prospects for continued economic growth in China, Taiwan and South Korea and economic recovery in Japan by raising the cost of imported raw materials. With China's economic growth increasingly serving as a major source of growth not only for the other Northeast Asian economies but also for the United States, Europe and Australasia, a terrorist attack on shipping in the chokepoints of Southeast Asia could have negative repercussions for the world economy as a whole.

However legitimate these concerns about the threat posed to maritime security by non-state actors may be, they should not obscure the fact that with the end of the Cold War the threat of disruption to maritime trade routes resulting from interstate conflict has not entirely evaporated. Indeed, some strategic analysts would argue that since the end of the Cold War the potential for interstate conflict at sea is increasing, and that sea power is becoming more, rather than less important in interstate relations (Friedman 2001: 1). During the March 1996 confrontation between China and the United States in the Taiwan Strait, when the People's Liberation Army (PLA) carried out exercises and missile tests near Taiwan, the US Navy, in the most significant display of American naval power since the 1950s, deployed two aircraft carrier battle groups, centred on the USS *Independence* and the USS *Nimitz*, to the vicinity of Taiwan. This was, as Robert Ross (2000: 87) describes it, 'a critical turning in post-Cold War US–China relations and in the development of a new regional order'. It was also a salutary reminder that great-power war in Asia, while far from inevitable, remains thinkable (Lim 2003: 1). And war, as E. H. Carr (1946: 109) reminds us, is always lurking in the background of international politics.

The 1996 crisis prompted strategic planners in both Beijing and Washington to adjust their defence policies to take account of the real possibility of a Sino-US war (Ross 2002: 48). Strategic planning for both sides is now based on the assumption that Chinese use of force against Taiwan will lead to war between China and the United States. In the aftermath of the 1996 confrontation, China initiated a series of measures to acquire the capability its armed forces would need to deny the United States Navy access to waters adjacent to the Chinese coast and surrounding Taiwan. Prominent among the platforms whose acquisition is designed to provide Beijing with such capability are the *Sovremenny*-class destroyers, the SU-27 and SU-30 combat aircraft, and the *Kilo*-class diesel-electric submarines. For its part, the crisis prompted Washington to increase arms sales to Taiwan, to consider a defence relationship with Taipei based on wartime cooperation and to take a greater interest in missile defence (Ross 2002: 48). According to Ross (2000: 117):

> The 1996 confrontation focused the Pentagon's attention on the US–China conflict over Taiwan as the most likely source of US involvement in a major war. Since then, planning for a war with China has become a Pentagon priority, with implications for budgets and weapons acquisition. Pentagon and Congressional interest in theatre missile defence, including cooperation with Taiwan on this system, has to a significant degree been a reaction to China's March 1996 missile tests.

The 1996 crisis highlighted the fact that in Northeast Asia, the principal theatre since the end of the Cold War for great-power rivalry, a contest for control of the seas is underway which, in structure if not yet in scale, resembles that of the great-power competitions for naval supremacy of the twentieth century: between Japan and Russia, Germany and Great Britain, Japan and the United States and the Soviet Union and the United States. Twice during the last century competition for predominance in East Asia was decided at sea: at the battle of Tsushima in May 1905, and in the great World War II sea battles of the Coral Sea and Midway in May and June 1942, and the Marianas, Leyte Gulf and the Philippines Sea in June 1944. It was this latter series of sea battles that laid the foundations for the American-dominated and -structured international order which prevails in the Asia-Pacific region to this day.

Unlike these earlier contests, however, the naval dimension of the strategic rivalry between China and the United States for pre-eminence in maritime East Asia is being played out increasingly below the surface of the East Asian seas. The symbols of this competition are not the imposing surface combatants of earlier naval contests – the battleships, cruisers and aircraft carriers – but the much less visible tactical submarines of the PLA Navy and the United States Navy's Pacific Fleet. Indeed, the hidden, silent contest which has begun to play out under the waters of East Asia may indeed prove John Keegan's (1993: 274) contention that we have now entered 'the era of the submarine as the predominant weapon of power at sea ... the ultimate capital ship, deploying the means to destroy any surface fleet that enters its zone of operations'.

Strategists from the Chinese Navy Research Institute in Beijing certainly seem to share this belief: 'the submarine will rise in status to become a major naval warfare force' (Shen *et al.* 1998: 263). The PLA Navy is accordingly equipping itself with a large, modern and effective submarine fleet in the hope of being able to exploit perceived weaknesses in the United States Navy's capabilities, particularly its vulnerable supply lines and its deficiencies in anti-submarine warfare (ASW) capabilities.

This study attempts to elucidate the strategic reasoning behind Beijing's quest for sea power and, in particular, its decision to give priority among the different instruments of naval power to the development of a strong tactical submarine fleet. In his 1976 essay on 'Sea Power and Political Influence', Hedley Bull (1976: 1) wrote that there were three questions worth asking about sea power:

> First, what political purposes does sea power serve? Why do nations seek to exercise military power at sea? Secondly, how does sea power promote these purposes? In what ways do nations use their sea power to further their political objectives? Thirdly, and more particularly – given that ours is an age in which the unlimited or unrestrained use of military power, at least between advanced states, is incapable of furthering political objectives, at sea or anywhere else – how can sea power be used as an instrument of foreign policy

in times of peace, or at all events in the absence of major war? What are the possibilities of 'gunboat diplomacy'?

According to the classic theory of sea power, navies have two purposes: the first is the protection of seaborne commerce in peacetime, or, in wartime, the protection of sea lines of communication or the denial of them to the enemy as a means of attack; the second, as Alfred Thayer Mahan (1965: 26) put it, is for 'aggressive purposes'. Bull (1976: 2) has described this aggressive purpose of a navy as bringing 'military power to bear in distant waters in order to support local clients, coerce local enemies, or neutralize one another's ability to act in the area'. In addition to the two purposes of sea power identified by Mahan, Bull (ibid.: 2–3) identified two more recent basic objectives underlying the exercise of sea power:

> First, some nations exercise sea power in order to deploy strategic weapons systems at sea or to take counter-measures against the strategic weapons systems of their opponents ... Countries wishing to deploy seaborne strategic deterrent forces need to shield them from detection, defend them from attack, and secure the ocean space in which they can move about. Countries threatened by seaborne deterrent forces are impelled to seek the means of detecting and destroying them ... The second objective ... is to seek to exert military power at sea in order to acquire or enlarge their share of the sea's resources, or protect it against threats from others.

In most respects, contemporary China's use of sea power accords with these basic purposes. And historically, for at least the past millennium, China has used naval power to protect commerce. For a brief period during the early Ming dynasty (1368–1644), China also used sea power to extend China's commercial and cultural influence beyond its immediate periphery. From the time of the Tang dynasty (618–907) onwards, when the southern, south-western and south-eastern coastal regions were absorbed into the northern Chinese political and cultural realm, the protection of commerce from smugglers and pirates became a constant preoccupation for imperial naval forces: it remains one for the PLA Navy. And as contemporary China becomes increasingly dependent on imported raw materials and energy for its rapidly growing economy, its interest in ensuring the protection of its sea lines of communication is growing commensurately.

The PRC's efforts to stake out and defend its offshore sovereignties and resources, which began immediately after its founding in 1949, are evidence that contemporary Chinese behaviour is consistent with Bull's description of the more recent uses of sea power. In 1950, for example, the PRC declared a fisheries conservation zone parallel to its coast in the Yellow Sea and East China Sea. And in 1951, Beijing, reiterating claims made by the defeated

Nationalist government, asserted that any disposition made at the San Francisco Peace Conference which did not recognise Chinese sovereignty over the Paracel and Spratly Islands would not be valid (Austin 1988: 44). Liu Huaqing, the 'father' of the modern Chinese Navy, summed up the importance of this aspect of sea power in an address to the China Military Science Association:

> Since the beginning of the 1970s, the strategic importance of the oceans has increased day by day. Exploitation of the ocean has turned into an important condition for coastal countries in developing their economy and overall strength of national power. It is certain that the ocean will be more and more significant to the long-term development of the country. We must understand the ocean from a strategic level and its importance to the whole nation.
>
> (Kane 2002: 64)

The development of a seaborne strategic nuclear deterrent, which the PRC began in July 1958 when Mao Zedong gave the green light to Project 09 – for the research and development of a nuclear-powered ballistic missile submarine (SSBN) – is evidence that China's quest for sea power is motivated by the first more recent reason given by Bull for states to pursue sea power (Lewis and Xue 1994: 6). The successful launch in 1988 of the *Ju Lang* (JL-1) intermediate range nuclear missile from its *Xia*-class (09) SSBN marked the fruition of China's efforts to deploy a seaborne nuclear deterrent (Cole 2001: 27).

However, until the arrival of the Europeans, very rarely in Chinese history had sea power been used by or against China for 'aggressive purposes'. For the greater part of its long history as a unified state, as John Fairbank (1974: 25) put it, 'Chinese naval power in the modern sense of the term remained abortive.' Chinese history has no record of naval expeditions by indigenous Chinese rulers such as, for instance, those mounted by the eleventh-century Chola King Rajendra I of Tajore in southern India against Sri Lanka and Southeast Asia (Keay 2000: 221). Chinese rulers did use sea power for aggressive purposes during the dynasties of the Mongol Yuan and the Manchu Qing dynasties. In the thirteenth century, during the brief reign of the Yuan dynasty (1239–1368), the Mongol conquerors of China mounted amphibious expeditions against Japan in 1274 and in 1281 (unsuccessfully), against Annam (Vietnam) between 1283 and 1288, and Java in 1292 (Swanson 1982: 25). And, in events prefiguring the present confrontation across the Taiwan Strait, the Qing dynasty also used naval power to overcome the remaining Ming loyalists who, under the leadership of Koxinga, had taken refuge on the island of Formosa. Qing forces took Formosa from the Ming loyalist forces under Koxinga's son in 1683, completing the Qing conquest of all China (Hsü 1995: 28). But these examples of the offensive use of Chinese naval power are notable for being so few and so exceptional in China's long history and extensive experience of warfare.

Sea power was used aggressively against China – or at least a Chinese tributary state – in 1592 and 1597 when, during the reign of the Wan-li emperor and the Japanese Shogun Hideyoshi, Japanese forces landed in Korea with the intention of conquering the Ming Empire. The Japanese were repulsed by Korean and Chinese troops and forced to evacuate the peninsula. It was not for another three centuries (1894) that Japan would mount another such amphibious invasion of Korea and dislodge the Qing dynasty troops from an area of traditional Chinese hegemony (Grousset 2000: 269). This Chinese defeat on land was accompanied by no less humiliating Japanese naval victories over the Chinese Peiyang fleet at the battle of the Yalu River in the Yellow Sea in September 1894, and at the Chinese naval base of Wei-hai-wei in February 1895 (Hsü 1995: 340). As a consequence of these defeats China was forced to cede Taiwan to Japan under the Treaty of Shimonoseki, thereby erecting an enduring obstacle to the development of sea power as a strategic instrument for the Chinese state. European naval power was, of course, instrumental in extracting the commercial and territorial concessions from the Qing Empire under the so-called 'unequal treaties', a process initiated by the British during the Opium War of 1840–1. France followed the British example by using its naval power to force Peking to open further ports to trade in the Second China War, which ended in 1859. The French Navy inflicted a further stinging defeat on Chinese naval forces at Fouzhou in 1884 (Swanson 1982: 80). According to one Chinese estimate cited by Srikanth Kondapalli (2001: xix), from the time of the Opium Wars until the founding of the PRC in 1949, foreign countries invaded the Chinese coastline more than 470 times.

China's response to the sixteenth-century Japanese invasion of Korea was consistent with the traditional Chinese response to threats to her maritime periphery. This was not to build and maintain a naval capability which could counter the threat offshore or attack the hostile forces at source, but rather to build coastal defences and to attempt to defeat the hostile forces once they had landed. An example of this defensive land-based strategy against maritime threats was the Ming military response to Japanese and Chinese seaborne raiders based in the islands of Japan and the Ryukyus – the East Asian equivalent of the Portuguese and English pirates who flourished in the Atlantic and the Caribbean and the Barbary pirates in the Mediterranean in the sixteenth century. As John Fairbank (1974: 23) notes, Ming China's defensive response to these raiders was closely modelled on China's traditional strategy for defending against the Xiong-nu raiders who emerged from the northern steppes and deserts, striking unpredictably to pillage and loot but not to seize and hold territory:

The Chinese response in the 1550s was even more defensive than it had been originally on the Great Wall. The resort was not to naval power, a counterforce at sea, but rather to guard posts, lookouts, and beacons on the coast backed by small garrisons and supplemented by special forces brought in to converge upon the pirates once they had pushed inland.

This pattern of land-based defence against maritime threats has persisted into the present era in the form of Mao Zedong's doctrine of people's war which held that intruders from the sea should be eliminated on land (Tien 1992: 278).

The concept of 'sea power' itself may not be an exclusively Western concept: Hervé Coutau-Bégarie (2000: 579) notes that Joseph Needham refers to the seventeenth-century author Zhang Xie, who employed the notion of 'command of the waters of the ocean'. But the whole of sea power theory is built on narrow historical foundations. The two main pillars of maritime strategic theory, Alfred Thayer Mahan and Sir Julian Corbett, drew on European history, and more particularly on British naval history, as the primary substantive basis for deriving their general principles. David Kang (2003: 58) has rightly sounded a note of caution about the pitfalls in seeking to use concepts and theories of international relations derived almost exclusively from European historical experience to explain interstate behaviour in Asia. The concepts and analytical models of maritime strategic theory should also be applied to Asia with due respect to the differences in European and Asian historical experience of the sea as a substantial factor in hegemonic warfare. One of the most striking differences between Asian and European military and political history is that the sea played a relatively small role in the expansion of Asian empires compared with its role in the evolution and expansion of successive European powers. This is particularly evident in the waxing and waning over the centuries of Chinese hegemony in Asia. As John Fairbank (1974: 3) has observed:

Considered geographically, the germ of Western expansiveness came from Graeco-Roman use of the sea, which fostered maritime trade, colonies and empires in the Mediterranean two thousand years before nineteenth century European imperialism, continuing in the same style, briefly took over the world. In contrast, the germ of China's defensiveness, her primary concern for social order at home instead of expansion abroad, came from her landlocked situation in North China remote from other centers of civilization and from sea routes communicating with them. Ancient China had no counterpart of the Cretan thalassocracy, the Trojan War, the Athenian navy, or Phoenician sea trade. The unified Chinese empire went through a full millennium of growth and change before the sea trade with nearby Southeast Asia became of any importance to it.

Of course, just because a nation's strategic culture is more or less exclusively that of a land power does not prevent it from taking to the sea when strategic conditions permit in an attempt to achieve whatever political goals it may have set itself. The United States, which until the 1890s had little or no tradition of sea power, is a case in point. And, as Colin Gray (1992b: 144) points out:

> It can be easy to forget that although Henry VIII constructed a fine navy, and some elements of an English naval policy can be traced back to the twelfth century, the fact remains that England in the sixteenth century was not heir to a great seagoing tradition. And earlier examples abound of land powers which have bucked their strategic traditions in order to make use of the sea to fulfil their ambition, from Sparta in the fifth century BC and Rome in the third century BC, through to the Arabs in the seventh and eighth centuries, and Germany and the Soviet Union in the twentieth century.

Nonetheless, it remains true that it was only in the nineteenth and twentieth centuries that China's experience of the use of sea power for 'aggressive' purposes became much more extensive than in previous centuries. For the most part, though, from the Opium Wars onwards China was the victim rather than the agent of the aggressive use of sea power. Although the PRC has used sea power aggressively on a number of occasions against Taiwan and nationalist-held islands in the Taiwan Strait, it has used its naval forces against other states on only three occasions: in 1974 to take possession of islands in the South China Sea (against South Vietnam), in 1988 (against Vietnam) and in 1995 (against the Philippines). China's historical experience with the use of sea power for 'aggressive' purposes, both as victim and as aggressor, raises the question of whether in the future the pattern of China's use of sea power will continue to be governed by predominantly deterrence and defensive objectives or whether its actions in the South China Sea herald a more offensive use of naval force.

It is with these general questions about the political purposes and uses of sea power in mind that we try to answer the more particular question of why its tactical submarine arm weighs so heavily in the overall balance of the force structure of the PLA Navy. In terms of the twin challenge that confronts any analysis of a state's armed forces, that of capabilities and intentions – the hardware and the software components of a military system – this study focuses deliberately on the intentions side of the equation, as the more elusive and nebulous of the two facets. It is, in effect, an attempt to understand – to borrow Admiral J. C. Wylie's (1967: 6) phrase – 'the patterns of thought of the military minds' of the decision-makers who are behind China's strategic choice to wager so heavily on its navy's tactical submarine arm. What reasoning lies behind Beijing's evident conclusion that submarines represent an important military means for attaining its political ends? To what use do China's strategic planners intend to put the PLA Navy's submarines? Strategically and operationally, how might China's submarine forces be employed in a conflict at sea?

At its broadest, the subject of this study is sea power. Its focus, however, is on submarines and undersea warfare because this seems to be the most salient manifestation of contemporary China's quest for sea power. Singling out a sole category of warship and one dimension of naval warfare in this way carries the risk of distorting the nature of sea power which, as Bernard Brodie (1944: 2) pointed out, 'has never meant merely warships. It has always meant the sum total of those weapons, installations, and geographical circumstances which enable a nation to control transportation over the seas during wartime.' It also risks creating the impression that submarines operate in isolation from other elements of a nation's naval forces, and indeed from its armed forces taken as a whole. But submarines constitute only one component of an integrated system of capabilities designed to produce tactical, operational and strategic effect in every domain of conflict including the sea surface, air, land, space, the electro-magnetic sphere and, as Chinese strategic thinkers from Sun Zi to Qiao Liang and Wang Xiangsu (2003: 212) as well as deterrence theorists have underscored, the human mind.

Whether or not we agree with Peter Padfield's (2000: 1) claim that 'maritime strategy is the key which unlocks most, if not all, large questions of modern history', the exploration of the PRC's efforts to develop a capability which will enable it to challenge the United States' maritime supremacy in East Asia allows us to examine Sino-US relations from an unconventional angle. An analysis of China's reasons for making its submarine fleet such an important component of its naval force structure also provides novel insights into Chinese naval strategy and provides potentially useful new information to help us to assess the risks of conflict in the Taiwan Strait. As Bernard Brodie (1959: 361) has remarked, 'strategy in peacetime is expressed largely in choices among weapons systems'. A close examination of Chinese strategists' choice to make the submarine a particularly important component of the PRC's naval force structure can therefore reveal much about the wider strategic perceptions and objectives of China's military and political leaders. Moreover, as Thomas Schelling (1966: 234) notes, while arms and military organisations may not necessarily be determining factors in international conflict, neither can they be considered neutral. Although originally conceived to play primarily a defensive role in naval operations, the submarine has more often been the instrument of choice for offensive operations by inferior navies. And possession of advanced, offensive weapons has often provided weaker states with the confidence to launch asymmetric wars against stronger opponents. Moreover, as Owen Cote (2003: 89) observes, historically, stronger navies have tended to underestimate the submarine threat to their sea lines of communication. The PRC's increasingly capable undersea warfare forces could therefore be a critical and potentially destabilising factor in the management of any future Taiwan Strait crisis, perhaps making all the difference between a peaceful political solution and a descent into war.

Although there is no lack of published assessments of China's strategic policy and military capabilities, and of the prospects for conflict in the Taiwan Strait, relatively little work has been published recently which focuses specifically on the PLA Navy and the naval dimensions of a Taiwan conflict – with the recent notable exceptions of Srikanth Kondapalli's *China's Naval Power* and Bernard Cole's *The Great Wall at Sea*. Given that a military confrontation between China, a continental power, and the United States, a maritime power, over the status of Taiwan is widely considered to be among the most plausible scenarios for the outbreak of war between major powers during the coming decades, and that in such a war 'the sea', as Corbett (1911: 15) put it, must necessarily be 'a substantial factor', it is puzzling that China's maritime strategy has not attracted more analytical attention than it has. This may be because ground force analysts tend to dominate the field of PLA studies (Stokes 1999: 1).[1] Or it may be because the naval dimension, although clearly critical given the environment in which a Taiwan conflict would take place, has been overshadowed by the rapid build-up by China of its ballistic missile forces in the Taiwan Strait theatre.

Open-source assessments focusing specifically on the PLA Navy's submarine fleet and the role that it might play in Taiwan Strait conflict are fewer still than more general surveys of China's naval capabilities. There are nonetheless signs of a quickening interest in the significance of China's increasingly capable submarine fleet for the management of security relations in East Asia. Two important articles were published on the subject in 2004 (Glosny 2004; Goldstein and Murray 2004). The authors of one of these articles, Lyle Goldstein and William Murray (2004: 164), conclude that 'a dramatic shift in Chinese underwater aspirations and capabilities is under way, and that submarines are emerging as the centrepiece of an evolving Chinese quest to control the East Asian littoral'.

Submarines are clearly very important to Chinese strategists, as this study tries to show. Obvious though it may seem, the major role played by submarines in China's naval strategy may have not have attracted the attention it deserves from Western analysts simply because submarines, by their very nature, are less visible than missiles, surface vessels or aircraft – a point often made by those who argue that there is no substitute for an aircraft carrier when it comes to manifesting naval 'presence'. As Michael Pillsbury (2001: 4) noted in outlining the gaps in Western analysts' knowledge of Chinese military affairs, 'as might be predicted, we are less knowledgeable about things that are less visible or tangible – training, logistics, doctrine, command and control, special operations, mine warfare – than we are about airplanes and surface ships'.

Analytical methods

The attempt to reconstruct the reasoning, and to capture at least some of the multiple considerations which combine to produce the strategic choice made by the PRC's political and military leadership to develop its tactical submarine fleet, relies on a methodological device pioneered by Graham Allison and Philip Zelikow in their seminal analysis of the Cuban missile crisis, *The Essence of Decision* (1999). In this study, Allison and Zelikow tried to develop a richer, more finely graded explanation of the decisions and actions of the central participants in the crisis than had hitherto been available. They did this by examining the events of October 1962 through three distinct frames of reference or conceptual lenses: the rational-actor model; the organisational behaviour model and the government politics model.

The rational-actor model, which is the dominant and usually implicit analytical model for explaining and predicting foreign and military actions, seeks to explain decisions and policies as the outcome of a rational – that is, purposive, goal-directed – response by unitary actors to a defined situation. Rational-actor model explanations of political or strategic behaviour are characterised by a cluster of assumptions: that the actor is a national government, and that the action is chosen as a calculated solution to a strategic problem. 'The explanation consists of showing what goal the government was pursuing when it acted and how the action was a reasonable choice, given the nation's objectives' (ibid.: 15).

The organisational behaviour model 'emphasizes the distinctive logic, capacities, culture, and procedures of the large organizations that constitute a government' (ibid.: 5). This model conceives of decisions and actions as not so much the product of rational, value-maximising choices of unitary actors but more as '*outputs* of large organizations functioning according to standard patterns of behaviour' (ibid.: 143). Explanations based on the organisational behaviour model proceed according to certain patterns of inference: 'if organizations produced an output of a certain kind at a certain time that behaviour resulted from existing organizational structures, procedures and repertoires' (ibid.: 6).

The government politics model seeks to explain decisions and actions in terms of the outcome of a bargaining process among key players in a national government.

To try to understand the reasoning behind the choice that has clearly been made by Chinese political and military leaders to create and maintain a naval force structure heavily biased towards tactical submarines, a combination of the first two conceptual models proposed by Allison and Zelikow promises a more complete explanation than one arrived at by relying solely on a classic rational-actor analysis. An analysis based on a governmental politics conceptual model would no doubt produce a still more complete explanation. However, this model is only capable of enriching explanations of military and political choices when the information about the conflicting ideas and interests which form the basis of the inter-agency bargaining process is relatively accessible – as it is in the case of the United States and some other Western democracies where a cornucopia of open-source material is available to provide insights into debates on military strategies, capabilities and force structures. In a closed, opaque, authoritarian political system such as China's, where secrecy about political and military affairs is culturally ingrained, it is much more difficult to obtain insights into the bargaining process among the competing civil, military and political factions – even for the sinologists and experts in Chinese military affairs

This study is therefore divided into two parts. The first part – from Chapter 1 to Chapter 8 – attempts to use a rational-actor type analysis to explain the PRC's strategic choice to develop a strong undersea warfare capability. It takes the approach of explaining China's submarine-heavy naval force structure as the rational choice that any reasonable strategist would make in the circumstances that Chinese strategists find themselves in, taking into account such factors as China's national political and military goals, its economic, industrial and military resources, military technology, its strategic geography, and judgements about the capabilities and strategies of its potential adversaries. This analysis draws particularly on the insights of theorists of maritime strategy in order to 'uncover the universal logic that conditions all forms of war as well as the adversarial dealings of nations in peacetime' (Luttwak 1987: xi). It thus implicitly accepts the premise that the implementation of state policy by armed force is divided into constant and variable components, the constant elements reflecting the permanency of human nature and physical geography, and the variable factors reflecting history, geopolitical circumstances, economics, and other unique influences on the strategic choices of a particular security community.

The second part of the study – from Chapter 9 to Chapter 11 – therefore approaches the issue through an organisational behaviour model of analysis which highlights the more subjective dimensions of the Chinese predisposition to a naval force structure geared towards undersea warfare. Implicit in this approach is that the many individuals and organisations which contribute to the decisions which collectively shape the PLA Navy's force structure – from regional fleet commanders through the PLA Navy Headquarters, Political, Logistics and Equipment Departments in Beijing to the Central Military Commission and the CCP Central Committee Politburo – constitute an 'organisation' whose decisions and actions are susceptible to an analysis using this model. In terms of their purposes and the results of their activities, the strategists, force planners and the political and military hierarchy responsible for the decisions which eventually determine the PLA Navy's force structure can be considered to constitute an organisation as defined by Allison and Zelikow (ibid.: 145).

The principal advantage of supplementing a rational-actor type analysis of decision-making with an organisational behaviour approach is that it enables the explanation to capture some of the more subjective factors peculiar to individual security communities which contribute to the final outcome of the decision-making process – in this case the PLA Navy's heavily submarine-biased force structure. For example, as Allison and Zelikow note, one of the characteristics of organisations is that they are oriented towards doing what they already do. Thus one factor favouring the maintenance of a strong tactical submarine fleet may simply be inertia: since its foundation, the PLA Navy has always had a numerically strong submarine arm; it is therefore organisationally disposed to maintaining this capability.

Another advantage of an organisational behaviour type of analysis, particularly for cross-cultural studies, is that it helps to overcome the inherent ethnocentrism of the rational-actor type analysis. Culture is an important element of the organisational behaviour model of analysis – an element that is seldom taken into account in rational-actor type analyses. The basis of a rational-actor model of analysis is the assumption of the decision-maker as a universal, abstract 'rational man' who behaves in a goal-directed, value-maximising way. This enables the analyst to proceed by projecting himself into the decision-maker's situation in order to explain the latter's decisions and actions as the rational choices of anyone in the same circumstances, with the same priorities, objectives and constraints. In his landmark study *Strategy and Ethnocentrism*, Ken Booth (1979: 65) explains the potential pitfalls of this approach for strategic analysts:

> When thinking about the rational behaviour of others, strategists tend to project their own cultural values. The habitual assumption is: 'If my opponent is rational, he will do what any other rational man would do in his situation. I am rational. Therefore he will do what I would do in his shoes. If I were in his shoes I would ...' This is how most strategic thinking proceeds, but it should be apparent that one can only predict the behaviour of a 'rational man' if both observer and observed share the same values, have the same set of priorities, and have similar logical powers. Ethnocentric perceptions interfere with this process: they mean that one's own values and sense of priorities are projected onto the other. By this process

ethnocentrism undermines the central act in strategy, that of esti-
mating how others see the world and then will think and act.

This study therefore attempts to mitigate the inevitable risk of ethnocentrism by examining some of the general beliefs, attitudes and behavioural
patterns which spring from the particular historical experience of China's
strategic policy community and which fundamentally shape Chinese strategic
thinking and policy decisions on issues such as military force structures and
doctrine. The study thus explicitly accepts the premise that culture and style
are useful keys for improving our understanding of why particular security
communities behave as they do, and the contention advanced by Colin Gray
(1986: 37) in his pioneering study on the influence of strategic culture on
American and Soviet nuclear strategy, namely that:

> major streams of policy decisions ... cannot simply be explained in
> terms of the characteristics of particular people, their unique
> assessment of policy options, and the bureaucratic-political milieux
> in which they find themselves – though very often these factors
> could (and did) help shape the mix of contending bureaucratic-
> political forces. In addition, it is necessary to consider the strategic
> culture of the policymakers.

The second part of this study therefore attempts to correct the inherent
bias in the rational-actor model of analysis adopted in the first part by
examining some of the more subjective factors particular to Chinese
strategic culture which may have a bearing on the choice of Chinese strategic
planners to emphasise undersea warfare capabilities in the composition of
their naval forces. It then attempts to explore some of the consequences of
this choice of instrument for the pursuit of Chinese sea power in terms of
the risk of a future Taiwan Strait crisis degenerating into war at sea between
China and the United States.

1

CHINA'S TACTICAL
SUBMARINE FLEET

The preponderance of tactical submarines in the PLA Navy's force structure
is testimony to the importance that Beijing attaches to maintaining an offensive undersea warfare capability. With 67 tactical boats, the PLA Navy's
submarine fleet currently outnumbers its 63 principal surface combatants
(International Institute for Strategic Studies [IISS] 2003: 153).[1] Among
other navies operating in the Asia-Pacific region, only North Korea's, with
26 submarines to three principal surface combatants, places proportionally
greater emphasis on undersea warfare capabilities (IISS 2003: 160). While
the Russian Navy as a whole has more tactical submarines than principal
surface combatants in its order of battle, its Pacific fleet has a ratio of only
five tactical submarines to eight principal surface combatants (IISS 2003: 92).

Although the Chinese submarine fleet's capability is greatly reduced by
the advanced age of most of its vessels, the upgrading of this capability is
also one of the highest priorities of the PRC's overall military programme.
This priority is evident in the US$1.6 billion contract that the PRC
concluded with Russia in May 2002 for the purchase of eight new *Kilo*-class
submarines (Pomfret 2002). Construction of these eight new 636 *Kilos*
began in March 2003 (Kanwa 2003b). Delivery is expected before 2007
(Cabestan 2003: 38).

The very size of the Chinese tactical submarine fleet, with 62 conventionally powered tactical submarines (SSK) and five *Han*-class nuclear-powered
attack boats (SSN), makes it one of the most formidable submarine fleets in
Asia, despite the obsolescence of a large proportion of its vessels (the technology of the *Ming* class goes back to the 1950s and that of the *Romeos* to
the 1960s). In quantitative terms at least, the South Korean fleet with 20
attack boats, Japan with 16 and Russia with 35 are clearly outclassed. Only
the American submarine fleet, with 54 nuclear attack boats (of which
around 27 are deployed to the Pacific Fleet) and 18 SSBNs, is numerically of
a similar order (IISS 2003: 26). The eight new project 636 *Kilo*-class
submarines – the best diesel-electric submarines yet produced by Russian
naval shipyards – will join two older project 636 *Kilos* and two less capable
export version 877 *Kilos* in the Chinese submarine fleet. With these 12

imported submarines, the 30 aging *Romeos*, 20 *Ming*-class submarines (first introduced into the fleet in 1969), the three more recent *Song*-class boats and its single SSBN, the PRC will be by far the world's largest submarine operator. According to the latest Pentagon assessment, by 2010 China will have withdrawn the *Romeo*-class submarines from service, and by 2020 China's non-nuclear submarine inventory will probably include *Ming*, *Song* and *Kilo* boats (United States Department of Defense [USDoD] 2003: 26).

In qualitative terms, however, the Chinese submarine fleet is currently outclassed by those of most other countries operating submarines in Northeast Asian waters. But the very fact that China is able to deploy such a large number of boats, even though many of them are old, noisy and slow, makes its submarine fleet a force to be reckoned with. Such a large fleet enables the PLA Navy submarines to patrol over a wide area. With an effective range of 11,000 to 12,000 kilometres, even for the oldest Chinese boats, in theory Beijing can deploy its submarine forces almost anywhere in the seas surrounding Japan and Taiwan, creating a significant challenge for potential adversaries to find and track them. It should be remembered that during the Second World War, Admiral Doenitz achieved remarkable success at the beginning of the submarine war in 1940 and 1941 with scarcely a score of operational U-boats (Couteau-Bégarie 2000: 602). It should also be remembered that during the Falklands War, just one well-operated, modern Argentine Type-209 diesel submarine, the *San Luis*, operating 800 nautical miles from its base, was able to make two attacks on British warships, and was capable of frustrating the capabilities of one of the most skilled ASW navies in the business (United States Naval Doctrine Command 1998).

With a dozen *Kilo*-class boats to its fleet, China will be able to present an even more serious threat to adversary navies operating in East Asian seas. The four *Kilo* submarines already in service are among the PLA Navy's most formidable assets. According to the Pentagon, they are one of the world's quietest diesel-electric submarines (USDoD 2003: 26). The 636 model incorporates an advanced engine design which gives it a longer submerged time and greater endurance without needing to recharge its batteries (Shambaugh 2002: 273). *Kilo*-class submarines are considered to be as capable as the advanced conventional boats of German, Dutch and Swedish design operated by other navies of the Asia-Pacific, including South Korea, Singapore and India (which has 10 *Kilo*-class boats), but less capable than the 16 Japanese submarines or the six Australian *Collins*-class boats (Holt 1999).

The *Kilos* are well armed, however, with weapons such as the Test-71ME heavyweight, wire-guided torpedo and the 53–65KE wake-homing torpedo. The new *Kilos* are likely to be equipped with the Klub weapons system – a powerful and flexible system which will enable them to fire land-attack cruise missiles, anti-ship cruise missiles with supersonic terminal homing, and rocket-thrown anti-surface and anti-submarine torpedoes (Goldstein and Murray 2003). The new *Kilos* will incorporate a number of improvements including better batteries, an improved digital sonar system, slower propellers and quieter engines. The conventionally powered *Kilo* submarines, with their hulls covered with anechoic tiles designed to defeat detection and targeting by active sonar, are as quiet as the American *Los Angeles*-class nuclear attack boats (Shambaugh 2002: 275). China's new *Kilos* may well incorporate an air-independent propulsion system, allowing them to stay submerged for up to 45 days (Cole 2001: 97).

The PLA Navy may, however, be experiencing problems in integrating its existing *Kilo*-class boats into its submarine fleet and in supporting their complex Russian high-technology electrical and combat systems (Bussert 2003). Two of the four *Kilos* which the PLA Navy currently operates have been seen at sea only infrequently (Shambaugh 2002: 273). Crew training has also been problematic and the boats have had to be returned to Russia for battery repairs (Cole 2001: 97).

China is also expanding and modernising its *Song*-class submarines, which will eventually replace its *Ming*-class boats. The *Song*, although indigenously designed and built, incorporates a mixture of Chinese and Western technologies. The first *Song*-class submarine was launched in 1994 and commissioned in 1996 (Shambaugh 2002: 272). The lead boat reportedly suffered from serious system design and operational problems caused by the integration of Chinese, Russian and imported systems such as the French TSM 2225 sonar and German diesel engines. It took several years to overcome these problems and to launch the second *Song* submarine (Bussert 2003). The *Song* is designed to carry the developmental YJ-82, China's first encapsulated anti-ship cruise missile capable of being launched from a submerged submarine (USDoD 2003: 26). It is likely that advanced Russian technology in the *Kilos* will find its way into future *Song*-class boats.

In July 2004, Western media reported the launching of a new generation of conventional diesel-electric submarines, named the *Yuan*-class by US intelligence. The programme is believed to have begun in 2002, with the first boat laid down at Wuhan Shipyard in late 2003 or early 2004. Details of this new submarine are sketchy, but some sources believe that it is comparable to the improved 636 *Kilos* in terms of size and general performance. Photographs of the new *Yuan*-class SSK suggest that it may have been derived from the *Song*-class and the *Kilo*-class boats (China Defense Today 2004).

In 2002, according to the Pentagon, the PRC laid down its first second-generation nuclear attack submarine of the Type-093 class. It believes the first Type-093 class SSN, which will also carry wire-guided and wake-homing torpedoes and cruise missiles, will become operational soon after 2005. Three more of this class will enter into operation by 2010. The Pentagon considers that the level technology of these boats approximates that of the second-generation Soviet *Victor III*-class SSN, and that they will eventually replace the PLA Navy's five first-generation *Han*-class SSNs, to spearhead China's capability to attack aircraft carriers (USDoD 2003: 27).

The PLA Navy also operates 10 submarine depot ships to support its submarine fleet at sea (IISS 2003: 154) and three *Dajiang*-class ships with a large crane and carrying two 35-ton deep submergence recovery vessels (DSRVs). It has built long-range high frequency (HF) and very low frequency (VLF) communication capabilities for supporting distant operations. Submerged communications have been possible since 1980 via VLF transmitters at Zhangian and Yulin (Bussert 2003).

The Chinese submarine fleet is divided into six flotillas of about 10 boats each. The Northern fleet has three flotillas, the Eastern fleet two and the Southern fleet one. The new *Kilos* have been allocated to the Eastern fleet. The Southern fleet will receive the recommissioned *Mings* when they enter into service. This suggests that the PLA Navy considers that the most capable submarines in its fleet will be best used in missions against Taiwan or in the South China Sea (McDevitt 2000).

The human factor of the Chinese submarine fleet is more difficult to assess than its material and technical capability. Fighting efficiency is more than just a function of the effectiveness of the weapon system: it is also determined by the experience, training, morale and military culture of the commander and crew. The PLA Navy has at least half a century's experience in submarine operations, longer than many of the navies operating submarines in the Asia-Pacific region. The accident suffered by the *Ming*-class submarine in March 2003, which cost the lives of some 70 crew members, does not necessarily reveal much about the quality of the equipment or crews of the Chinese submarine service. Submarine operations are naturally hazardous activities and even the most experienced, well-trained and well-equipped submarine services suffer occasional serious mishaps.

The ability of the PLA Navy to integrate new and sophisticated equipment into its fleet and to make effective use of new capabilities should not be underestimated. One of the conclusions reached by Larry Wortzel (2003: 68) from his study of the PLA's Beiping–Tianjin campaign of 1948–9 is that the Communist forces displayed a remarkable capacity to use effectively newly acquired, relatively sophisticated equipment obtained from defeated Nationalist forces, a lesson which he considers relevant to the contemporary PLA's integration of new, mainly Russian, equipment.

On the other hand, there are more intangible factors which may have an impact on the fighting efficiency of the PLA Navy's submarine force. Most submariners would contend that in the submarine service, more than any other branch of a navy, the personal qualities of individual commanders are a critical factor in the overall performance of the vessel. During the Cold War, Western naval analysts believed that the fighting efficiency of the Soviet Navy's submarine arm was undermined by the rigid command and control arrangements which flowed from the Soviet state's insistence on maintaining tight political control over its armed forces. As two former

Royal Navy submarine commanders (Moore and Compton-Hall 1987: 58) assert,

> Western submariners could not possible operate efficiently, offensively and successfully under such conditions. A commanding officer enjoys, and makes the best of, the freedom insisted upon by Doenitz for a U-boat captain: 'He is independent and master of his own decisions.'

The reform and modernisation initiatives undertaken over the past decade and a half are making China's armed forces increasingly professional. But the strongly hierarchical model of the PRC's Leninist state system of civil–military relations, superimposed on the hierarchical structure of China's traditional Confucian and legalist culture and society, fosters an ethos which is at odds with the individualism that is the hallmark of the successful submarine commander. As Captain Moore and Commander Compton-Hall succinctly put it (ibid.: 55): 'Good captains are individualists, and individualists are bad communists.'

More concretely, there are reports that the overall standards of training, maintenance and equipment operation in the PLA Navy are not very high (Holt 1999). If the frequency and conditions of training observed generally in the PLA are an indication of the standards of training in the submarine service of the PLA Navy, then they fall far short of those common in Western navies (Shambaugh 2002: 95). Beijing has reportedly skimped on training for the *Kilo* crews. When the original agreement for the supply of the 877 EKM *Kilo* submarines was negotiated in 1994, the Russians proposed 18 months of training, a realistic simulator and associated base support infrastructure. But Beijing agreed to fund only 12 months of training for the first 877 crew in St Petersburg and a shorter training period for the second crew at the submarine base in Xiangshan. In 1997 Beijing reduced the training period for the two 636 crews to nine months, and only the officers were trained in St Petersburg (Bussert 2003). Some naval units have reportedly been criticised for conducting night training at dusk and submarine training on the surface rather than submerged (Shambaugh 2002: 97). A shortage of trained crews means that Chinese submarines spend relatively little time at sea (Holt 1999). Moreover, Chinese submarine crews have no recent experience of wartime operations. Nonetheless, Shambaugh (2002: 102) notes that, during 1999, Western military intelligence reported that PLA Navy submarines from the East Sea fleet and the South Sea fleet 'were undertaking much longer sea patrols – forty-five days for diesel-electric and sixty days for nuclear-powered subs – and were doing so around the eastern coast of Taiwan for the first time'. Much of this submarine training reportedly '[employed] "confrontational training" sub-versus-sub techniques'.

Together with its efforts to upgrade its recruitment, training and doctrine, improvements to its submarine force, represented by the acquisition of the *Kilo*-class boats and their weapons systems and the continuing *Song*-class construction programme, are evidence of the great importance that the PLA Navy attaches to the development of an effective submarine warfare capability. Chinese military planners appear to agree with the conclusions of three PLA Navy officers from the Navy Research Institute in Beijing on the critical role of submarines in future warfare: 'during the First World War, the dominant vessel was the battleship, and in World War II it was the aircraft carrier. In future global wars, the most powerful weapon will be the submarine.' They conclude that 'it is an important aspect of navy restructuring to develop and maintain submarine forces' (Shen *et al.* 1998: 283). American naval officers also recognise the significance of China's emphasis on developing an effective submarine warfare capability. In the view of Rear Admiral Michael McDevitt (US Navy, retd) (2000): 'submarines are an essential ingredient in China's maritime strategy. Submarines are central to any attempt to blockade Taiwan, and are the best chance the PLA Navy has to delay or deal with the USN carrier battle group.' Goldstein and Murray (2003) maintain that 'perhaps the most significant development for the US Navy is China's extensive efforts to upgrade its submarine force' and warns that 'the US Navy should not underestimate China's ability to build a capable submarine force to challenge a superior foe'. The Pentagon's 2004 annual report on the military power of the PRC states that 'one of the top priorities for the PLAN during the 10th Five Year plan is the manufacturing of submarines' (USDoD 2004b: 39). Further evidence of the priority accorded by Beijing's military and political leaders to the development of the PRC's submarine fleet is the appointment in June 2003 of another submariner, Admiral Zhang Dingfa, as the Commander-in-Chief of the PLA Navy. This means that over the last 16 years, two out of three PLA Navy commanders have had a background in submarines. Admiral Zhang replaced Admiral Shi Yunsheng, a naval aviator, when the latter resigned following the accident in March 2003 to the *Ming*-class submarine. Shi's predecessor, however, Admiral Zhang Lianzhong, who succeeded Liu Huaqing in January 1988 and remained in the post until the end of 1996, was also a submariner (Kondapalli 2001: 197).

What then is the strategic logic behind Beijing's drive for an effective submarine warfare capability? And what strategic objectives do Chinese political and military elites hope to attain through the development of this capability?

2

THE GEOPOLITICAL CONTEXT

China – twenty-first-century *perturbateur?*

As a result of the collapse of the Soviet Union, for the first time for two centuries China no longer confronts a major threat to her security on her land borders. As Robert Ross (1999; 2000: 170) has observed, China was the principal strategic beneficiary in East Asia of the evaporation of Soviet power. Wherever there had been Soviet influence in a third country in Asia, it was replaced by Chinese influence, whether on the Korean Peninsula where Chinese domination replaced Sino-Soviet rivalry or in Indochina where Moscow's retreat enabled Beijing to re-establish its historic influence. This has resulted, according to Ross, in an East Asia dominated by two great powers, one continental and the other maritime: China dominates mainland East Asia while the United States dominates maritime East Asia. The only exception to this pattern is South Korea's alliance with the United States which – until the Pentagon's announced reduction of this number by a third over several years from the end of 2004 – has enabled Washington to station 36,000 troops on the East Asian continent (Rhem 2004).

From the perspective of sea power theory, the question that now arises is whether, having re-established its former dominance over continental East Asia, China is displaying the natural tendency of continental powers to want to transform their continental pre-eminence into maritime dominance (Gray 1995: 72). This theory would suggest that, at the beginning of the twenty-first century, China is in a position similar to that of France in 1793, Germany in 1905 and 1940, or the Soviet Union in 1945: pre-eminent continental powers in their respective regions, they found their aspirations to exercise uncontested regional hegemony frustrated by the great maritime powers of the era. Among the avowed enduring national interests of the United States is the preclusion of 'hostile domination … of the East Asian littoral' (USDoD 2001: 2). China's great ambition to re-establish its historic role as the pre-eminent power in East Asia can therefore only be realised by challenging the United States' command of East Asian seas. In positioning China to challenge America's pre-eminence in East Asia, Chinese leaders today know what Japan's leaders knew in 1941: that regardless of the ultimate magnitude of America's strength, it cannot be brought to bear in the western Pacific save through the instrumentality of sea power (Brodie 1944: 1).

China's avowed intention to construct a blue-water navy to enable it to become a global force by 2050, a long-term force-planning goal adopted in the late 1980s, recalls the goal adopted a century earlier by that other rising power, the United States. With the adoption of the Naval Bill in 1890, the American Republic abruptly reversed its longstanding defensive posture and adopted a programme to develop the means to give its foreign policy both the muscle and the global reach to enable it to become a world class power. The year 1890 'marked the onset of a revolution in doctrine which trans-formed the United States Navy from a loosely organised array of small coast defenders and light cruisers into a unified battle fleet of offensive capability' (Rhodes 1996: 80). The argument advanced in November 1889 by the US Secretary of the Navy, Benjamin Tracy, to justify the shift in the United States Navy's force structure from one designed for coastal defence to one of a blue-water battle fleet capable of seeking out and destroying opposing fleets, might equally have been made by the Chinese military leadership in the 1990s:

> We must have a force to raise blockades ... We must have a fleet of battleships that will beat off the enemy's fleet on its approach, for it is not to be tolerated that the United States ... is to submit to an attack on the threshold of its harbours. Finally, we must be able to divert the enemy's force from our coast by threatening his own, for a war, though defensive in principle, may be conducted most effec-tively by being offensive in its operations.
>
> (Rhodes 1996: 81)

It took America a quarter of a century to raise its navy from twelfth rank among world navies to second place. A century later, China has given itself half a century to realise a similar ambition. For many analysts, the strategic rivalry between China and the United States which has emerged since the end of the Cold War has become the principal dynamic of international politics in early twenty-first-century Asia. It has all the hallmarks of the major power confrontations which shaped European and global history during the last two centuries.[1] Indeed, the growing competition between the United States and China for pre-eminence in East Asia looks increasingly like the latest episode in a persistent world historical pattern of conflict between great land powers and coalitions organised by great offshore sea powers – a pattern which Colin Gray (1990: 60) considers to have been 'too steady to be dismissed as a passing phase or an accident of particular circumstances'.

The pattern has been for the dominant land power, once it has consoli-dated its position on the continent, to develop maritime strategic ambitions beyond the traditional strategies of coastal defence and *guerre de course* and instead construct an ocean-going 'great navy'. In his *Théories stratégiques*,[2] French Admiral Raoul Castex (1878–1968) – whom Gray includes in his select list of significant strategic thinkers of the twentieth century (1999b: 86) – calls these rising continental states which develop maritime ambitions '*perturbateurs*' – 'disturbers [of the peace]' (Castex 1997: vol. V, 129). Writing in the aftermath of the First World War, he no doubt had Germany foremost in his mind when he portrayed the general traits of these *perturbateurs*, but his description could equally well apply to China today:

> The *perturbateur*, at first threatening because of its terrestrial expansion, then becomes so because of its consequential naval and commercial development, which gives it the means to extend its reach to those nations which seemed until then beyond its hegemony ... This maritime expansion is a natural consequence of the rising movement which carries the *perturbateur* forward ... And when such vigour appears, it involves all domains at the same time: military, scientific, political, commercial, industrial, maritime, colonial, demographic. The excess of vitality is not limited to a single direction.

One notable difference between today's Sino-US great-power confrontation and the US–Soviet confrontation in the second half of the twentieth century, however, is that in the latter case, the central front was terrestrial, lying across the North German plains (with a subsidiary terrestrial front along the Sino-Soviet border in Central and East Asia). As a consequence, the naval component of US–Soviet strategic rivalry, while an essential enabling element of the military dimension of this conflict – especially after the early 1960s when each side developed the undersea legs of their strategic deterrent triads – nevertheless played a subsidiary role in the Cold War conflict. In contrast, the front line in the Sino-US confrontation is offshore and the maritime dimension plays a central role. This front line lies somewhere in the East and South China Seas between the Chinese coast and what Chinese military strategists term 'the first island chain' (the line of islands in the western Pacific running from the Japanese archipelago in the north, through the Ryukyu Islands and Taiwan, to the Philippine archipelago in the south). In today's Sino-US rivalry, sea power has a much greater prominence and centrality than it had in the US–Soviet confrontation. In the great-power confrontation of the latter part of the last century, the Soviet Union, as a continental power, had the advantage of confronting its principal rival, an alliance of maritime and continental states led by a predominantly maritime power, in its own element – on land. By contrast, China's geographic circumstances and historic traditions immediately put it at a disadvantage in its strategic competition with the United States.

Despite the fact that China has more than 17,000 kilometres of coastline and over 6,000 islands, history and geography have conspired to make China's strategic culture one of a continental rather than a maritime state. Although the extent of China's territory has waxed and waned through history, its current land borders are more than 16,000 kilometres in length. By way of comparison, the northern borders of the Roman Empire at the time of Augustus – from Spain in the west to Jerusalem in the east – measured some 8,000 kilometres (Swaine and Tellis 2000: 9). Thus, like France, Germany and Russia, China is a continental power whose geography has obliged it to divide its military resources between its land forces and its maritime forces, thereby precluding its development as a great sea power. Apart from the celebrated voyages of Zheng He between 1405 and 1435 as far as the East African coast (and perhaps beyond) (Menzies 2002), and the brief and unsuccessful efforts towards the end of the nineteenth century to build a navy to rival Japan's, China's imperial naval traditions are overshadowed by its overwhelming preoccupation with land warfare. As Bernard Cole (2001: 4) remarks: 'the navy was at various times capable and even powerful, but it was never vital to a dynasty's survival'. At the birth of the People's Republic in 1949 the People's Liberation Army Navy did not exist. The new regime had to build a navy to defend its coasts and islands from the attacks of the Kuomintang and their American allies. The PLA Navy was only formally established in May 1950 (Cole 2001: 10). The PLA Navy's submarine force was first organised in 1951. Its first base was established at Qingdao in 1952, together with a submarine school the following year. PLA Navy submarine operations began in June 1954 (Cole 2001: 79).

In 1953, Mao Zedong decreed the building of a powerful navy to protect the new People's Republic's coastline from the imperialist bullies: 'we must build the coastline ... into a Great Wall on the sea' (Kondapalli 2001: 6). But for the first three decades of its existence, the PRC had neither the time nor the resources to fulfil this ambition. The Korean War in 1950, the split with the Soviet Union in 1961, the war with India in 1962, the confrontation with the USSR from 1961 to 1985, and the wars in Indochina, forced Beijing to focus its attention and its military resources on the defence of its continental frontiers. As a result, until barely a decade ago the Chinese Navy has been the poor relative of the PRC's armed forces, equipped with a number of ancient vessels, and restricted to a coastal defence role (Brisset 2002: 93).

In 1979, at the time when he began the radical economic reforms which, in less than two decades, transformed China into the most dynamic economic power in Asia, Deng Xiaoping called for 'a strong navy with a modern combat capability'. This call was repeated by Jiang Zemin in 1997 who, echoing Mao's words, instructed the navy to 'build up the nation's great maritime wall' (Cole 2000: 279).

Mikhail Gorbachev's accession to power in Moscow in 1985 and the subsequent evaporation of the Soviet threat triggered a radical revision of

Beijing's strategic thinking. The expectation of a total nuclear war or a major armed conflict with the Soviet Union gave way to the belief in the probability of small-scale local conflicts in neighbouring countries aimed at defending China's borders or contested islands. This change in Beijing's strategic perspective had important consequences for the development of China's maritime strategy. For the first time since the founding of the PLA, the navy has become a major player in the operational situations which China's armed forces may one day have to confront, whether it be a conflict with Taiwan or Japan – conflicts in which Beijing could not afford to discount the possibility of intervention by the United States – or a conflict to defend Chinese sovereignty over contested maritime territories.

Drawing the analogy between the Soviet Union's challenge to the American command of the seas since the late 1960s and Imperial Germany's challenge to Britain's naval position after the enactment of the Navy Law in 1898, Hedley Bull (1976: 4) observed:

> In both cases the challenging power is prodded into this course by an outstanding and dedicated naval leader – in one case Admiral Gorshkov, in the other Admiral von Tirpitz – deeply fascinated by the history of the use of sea power by the Anglo-Saxon countries, displaying the same mixture of resentment and envy of their position, seeking to arouse his country to a sense of its naval destiny, and fired by Mahan's doctrine that sea power is a necessary condition of great-powerhood.

As with Imperial Germany and the Soviet Union, China's drive to develop a blue-water navy is identified particularly with one man, PLA General Liu Huaqing. As Commander-in-Chief of the PLA navy from 1982 until 1988, Liu was the chief architect of the new Chinese maritime strategy developed in response to the new post-1985 strategic and economic circumstances (Cole 2001: 165). Liu (born 1916) was a veteran of the Long March and the PLA's Second Field Army, whose political commissar was Deng Xiaoping (Swaine 1992: 23). The Second Field Army had a reputation for its professionalism within the PLA, and its veterans were among the foremost advocates of modernisation and professionalism of China's armed forces. Other key leaders of the Second Field Army included Liu Bocheng, a trenchant critic of Mao's military pedantry, and Peng Dehuai, commander of China's military forces in the Korean War, and a strong advocate of reform and modernisation of the PLA from 1953 until 1959 when he was purged because of his criticism of Mao at the Lushan Plenum. The PLA Navy high command came to be dominated by officers associated with the Second Field Army (Whitson and Chen-hsia 1973: 195).

Following the Communist victory over the Nationalists and the founding of the PRC in 1949, Liu became vice-president and deputy political commissar in the First Naval Academy from 1952 to 1954. He was then sent for training to the Voroshilov Naval War College and the Frunze Naval Academy in Leningrad, where one of his professors was Admiral Sergei Gorshkov (1910–88), who was subsequently appointed by Khrushchev in 1956 as Commander-in-Chief of the Soviet Navy. Liu was strongly influenced by Gorshkov's thinking and is reported to have advised his colleagues in the PLA Navy to read Gorshkov's book, *Sea Power of the State* (You 1999: 165). A central theme of Gorshkov's book, which would have struck a nationalist chord with Liu and others among the PRC's first-generation leaders, was that a strong fleet was a prerequisite for membership in the select club of great powers (Gorshkov 1979: 66). Liu Huaqing's appointment as the PLA Navy's Commander-in-Chief in 1982 is generally regarded as the turning point in the development of China's navy. His drive to modernise China's naval forces by equipping them with high-tech platforms and weapons and by revolutionising strategy, doctrine and training prompted some naval analysts to dub him 'China's Gorshkov' (You 1999: 164).

China's new maritime strategic orientation was both necessitated and reinforced by the economic take-off triggered by the post-Maoist policy of opening China to international trade and investment. This had the effect of moving China's economic centre of gravity from the agricultural and heavy industrial areas of the interior towards the towns and trade and production complexes in the coastal regions, which are more vulnerable to attack by enemy naval forces. At the same time, the opening of China's economy to outside forces increased its maritime dependence. From the 1980s onward, all the elements of maritime dependence grew strongly: maritime traffic, the merchant marine, naval construction and industry, fisheries and the importance of off-shore resource zones (Cable 1995: 106). Today, 50 per cent of China's economy depends on international trade, of which 90 per cent is transported by sea (Cole 2001: 297). More important still is China's growing dependency on imported oil. China's oil imports have risen 150 per cent over the past five years and reached 69 million tons in 2002. China became the world's number two importer in 2004. By 2030, according to the International Energy Agency, China's oil imports will exceed those of the United States (CNN 2003). Because the greatest proportion of these imports comes from the Persian Gulf, the protection of sea lines of communication, especially those which traverse the Indian Ocean, the Straits of Malacca and the South China Sea, is a growing concern for Chinese strategists.

It remains true, however, that the PLA is first and foremost a land army. The subordinate status and role of the PRC's naval forces are reflected in their designation as the 'People's Liberation *Army* Navy' – recalling the old post-revolutionary Soviet 'Naval Forces of the Red Army'. As Shambaugh (2002: 154) points out, 'despite considerable reductions and reorganizations in recent years, [ground forces] remain the dominant service in terms of manpower, resources, doctrine and prestige'. Even after the substantial changes in the membership of the Central Military Commission (CMC) following the 16th Congress of the Chinese Communist Party (CCP) in November 2002, the CMC is still dominated by senior officers of the PLA ground forces (Shambaugh 2003: 50). However, the expansion of the membership of the CMC following the resignation of Jiang Zemin as Chairman of the Commission in September 2004 to include the chiefs of the PLA Navy, Air Force and the Second Artillery (Strategic Missile Forces) is an indication that the army's near monopoly of the senior military hierarchy is breaking down (Kahn 2004: 1). Nevertheless, these changes are gradual. At the regional level, the army is still paramount. While the PLA Navy commander is of equivalent rank to the military region commanders (Cole 2001: 80), for example, the commanders of the three PLA Navy fleets are still subordinate to the commanders of their respective military regions (Shambaugh 2001: 165).

Still, according to You Ji's study of China's armed forces (1999: 170), 'in the last decade or so the PLA Navy has received unprecedented leadership attention, budgetary allocations and increased human and natural resources'. As a result of this rise in status,

> naval allocations have in recent years climbed close to one-third of China's defence budget. Similarly, its manpower strength has risen from 8 per cent to about 10 per cent of the PLA's total, in contrast to the reduction of the army from 81 per cent to 70 per cent.

The increasing importance of China's maritime strategic interests and the growing political and economic stakes for China's leadership in a satisfactory resolution of the Taiwan issue are reflected in the rise of the PLA Navy's status and the fact that it continues to be relatively successful in avoiding the large manpower reductions which have been imposed on the PLA's ground forces. In the round of reductions carried out between 1997 and 2000, for example, naval forces were reduced by only 11.6 per cent compared with the 19 per cent reduction in the number of ground forces (Shambaugh 2002: 153). The high priority that China's political and military elite give to the modernisation of the PLA Navy is evidence of their assessment that it is China's maritime forces rather than its land forces which will be in the front line in the defence of China's national interests in the coming decades. Further evidence in the importance attached by the PRC leadership to the modernisation of the PLA Navy was the review of a naval exercise in South China in October 1995 in which Jiang Zemin led all the CMC members. At the review each of the CMC members committed himself to supporting the development of the navy as an urgent priority (You 1999: 170).

The rising status of the PLA Navy may also be evident in the appointment of Vice Admiral Zhang Dingfa as President of the Academy of Military Sciences (AMS), the PLA's highest organ of research on military theory and doctrine, in the reshuffle of the PLA High Command by the 16th Party Congress. This was the first time that a naval officer had been appointed to head the AMS since its foundation in 1958, or indeed any leading PLA institution. Such appointments had hitherto been exclusively the preserve of ground force officers (Shambaugh 2002: 59). In June 2003, following the accident to the *Ming*-class submarine and the consequent resignation of Shi Yunsheng, Zhang was promoted to become Commander-in-Chief of the PLA Navy. The appointment in August 2004 of Vice Admiral Wu Shengli, former deputy commander of the Guangzhou Greater Military Region, as one of the Deputy Chiefs of Staff of the PLA General Staff Department (GSD), the largest and the most important of the four general headquarters of the PLA, may be another indication of the growing importance of the navy within the PLA. Wu was appointed to head the GSD, which has overall command authority and responsibilities within the PLA, together with a PLA Air Force officer, Lieutenant General Xu Qiliang. Their appointment marks a break from the tradition of appointing army officers to command the GSD (Ching 2004).

Asserting Chinese sovereignty – the strategic importance of Taiwan and other offshore territorial claims

In addition to its growing maritime dependence, China has another reason for pursuing its policy of modernising and expanding its naval forces: since the 1980s, Beijing has claimed sovereignty over the Spratly Islands, a low-lying system of islands and reefs in the South China Sea. In 1992, the Chinese Congress of People's Deputies passed its law on the Territorial Sea and Contiguous Zone claiming the South China Sea as sovereign waters (Austin 1988: 53). China now lays claim to an expanse of ocean covering some 3.6 million square kilometres, although two-thirds of its claimed territorial waters are subject to dispute (You 1999: 161).

China's interest in affirming its sovereignty over the Spratlys and the Paracel Islands further to the north can be explained in part by the possible existence of exploitable hydrocarbon and mineral reserves in the shallow waters of the South China Sea (Dénécé 1995: 258). But the importance attached by Beijing to asserting its claim to a maritime territory which encompasses virtually the whole of the South China Sea probably relates as much to the geostrategic significance of these waters. The fact that the most direct maritime routes between the Pacific and the Indian Oceans pass through the straits which issue out into the South China Sea endows this area with a strategic importance for all maritime powers, and a particular strategic significance for the industrialised economies of Northeast Asia, including China. Although the Spratlys and the Paracels offer no surface areas extensive enough to sustain large-scale military activities, they can be used as outposts in these navigationally hazardous and disputed waters.

China's maritime claims also extend to other strategic islands offshore from continental East Asia, including the Diaoyu/Senkaku Islands (occupied by Japan in 1910) and – of critical importance for the PRC's grand strategy and internal political order – Taiwan.

The restitution of the integrity of Chinese territory is at the heart of the PRC's grand strategy. As Leninist and Maoist ideology fade into irrelevance with the spectacular expansion of the Chinese economy, nationalism has replaced Communism as the force for the political and social cohesion of the Chinese state. Increasingly, the legitimacy of the CCP depends on its capacity to realise the nationalist ambition of wiping out once and for all the 'century of humiliation' which preceded the founding of the PRC. The restoration of Beijing's sovereignty over the territory ruled over by the Chinese Empire before the European and Japanese imperial powers began carving it up in the nineteenth century is central to the PRC's nationalist project. Since the restoration of Beijing's sovereignty over Hong Kong in 1997 and Macau in 1999, the reintegration of Taiwan into a single Chinese political entity has become the next stage of this quest to realise the nationalist dream. Moreover, the continued existence of Taiwan as a separate political entity from the PRC means that the Communist revolution of 1949 remains incomplete, and therefore the legitimacy of the CCP's rule over China is less than absolute. Thus, Chinese nationalism, the People's Republic and the legitimacy of the CCP's rule are tightly interwoven (Cabestan 2003: 30).

For Beijing, however, Taiwan is not only a political issue: it is equally an issue of great strategic importance. Taiwan is the greatest prize in a strategic competition between China and the United States for supremacy in Asia. As the Pentagon succinctly puts it in its 2002 report to Congress:

> Beijing assesses that the permanent separation of Taiwan from the mainland could serve as a strategic foothold for the United States. At the same time, securing control over Taiwan would allow the PRC to move its defensive perimeter further seaward.
>
> (USDoD 2002: 10)

Only 150 kilometres from the Chinese coast (roughly the same distance as between Cuba and the tip of the Florida Peninsula), the island of Taiwan contains a number of important ports and air bases which have been used in the past, and would be used again in any future Taiwan Strait conflict, to attack the mainland and to challenge the PLA Navy and Air Force for the control of China's littoral seas. The control of the island of Taiwan is thus crucial for the security of China's mainland territory. Japan's rise to become the dominant naval power in the East Asian seas at the end of the nineteenth century was due in large part to the acquisition of sovereignty over Taiwan and the adjacent Pescadores (Penghu) Islands after its victory over China in the 1894–5 Sino-Japanese war. Possession of these islands gave it 'a geographical stranglehold on the waters of the China Seas' (Lindberg and Todd 2002: 82). General Douglas MacArthur famously described Taiwan as 'an unsinkable aircraft carrier and submarine tender' (Nathan and Ross 1997: 206). It was from their bases on Formosa that Japanese planes first attacked American forces under MacArthur's command in the Philippines on 8 December 1941 (Spector 1985: 109). The island's strategic significance was one of the reasons why Mao Zedong instructed the PLA in June 1949 to prepare to seize Taiwan, arguing that 'if Taiwan is not liberated and the [Kuomintang's] naval and air bases not destroyed, Shanghai and other coastal areas will be menaced from time to time' (He 2003: 74).

However, Beijing's strategic stakes in the control of the island of Taiwan are not simply defensive nor just limited to its value as a platform for attacks against the mainland. Taiwan lies astride the maritime lines of communication which link Europe and the Middle East with China, Korea and Japan – among the world's busiest and most important to the global economy. Whoever controls Taiwan, therefore, has a potential stranglehold over the largest and most dynamic economies of the Asia-Pacific region, including that of the PRC itself.

Of perhaps even greater strategic importance for Beijing is the significance of Taiwan to the realisation of any long-term ambition that the PRC's leaders may have to see China supplant the United States as East Asia's pre-eminent power. If, as Condoleeza Rice maintained before she was appointed by President George W. Bush as National Security Advisor, 'China is not a "status quo" power, but one that would like to alter the balance of power in its own favor' (Rice 2000: 56), then gaining control of Taiwan becomes an unavoidable step towards the achievement of this goal.

To supplant the United States as the predominant power in the regional hierarchy, or to restore China to the position traditionally held by the Chinese Empire in the pre-colonial era as the 'centrepiece' of an Asian inter-state system (Terrill 2003: 254), would imply the possession by Beijing of a credible nuclear deterrent in relation to the continental United States. If one accepts that the international hierarchy is determined by the relative political power of the states in the system and, as Hans Morgenthau (1993: 31) put it, that 'in international politics … armed strength as a threat or a potentiality is the most important material factor making for the political power of a nation', then, at least since 1945, there has been no more potent means to guarantee state sovereignty and assert political power internationally than the possession of nuclear weapons. As Robert Gilpin (1981: 215) points out, the possession of nuclear weapons not only provides the nuclear state with 'an infrangible guarantee of its independence and physical integrity' but also 'largely determines a nation's rank in the hierarchy of international prestige'. Certainly for the PRC's political and military leaders, nuclear weapons have been integral to their quest to overcome the legacy of the 'century of humiliation' and to restore China to its traditional position as the leading state in

Asia. National self-esteem, national autonomy and security, regional pre-eminence and global influence were all important elements in Mao Zedong's decision to seek a Chinese nuclear deterrent. In the arena of international politics, China's nuclear arsenal is perhaps the most expressive practical demonstration of Mao's famous dictum that 'political power grows out of the barrel of a gun'.

Almost half a century since Mao Zedong's decision in January 1955 that China should build its own nuclear arsenal, and forty years since the first Chinese nuclear explosion on 16 October 1964 (Lewis and Xue 1988: 11), China has endowed itself with an arsenal which underpins a doctrine of 'minimum deterrence', whereby Chinese nuclear forces could be assured of absorbing an initial first strike while still being able to retaliate with inter-continental missiles into the homelands of the United States, Russia or India (Shambaugh 2002: 91). China's ability to pose a nuclear threat to the continental United States currently depends on around twenty DF-5 and DF-5A (CSS-4) intercontinental ballistic missiles, first deployed in 1981, which have a range of around 13,000 kilometres and each carry one 4 to 5Mt warhead. The Chinese nuclear arsenal also includes a range of theatre weapons which could be used to target US and allied forces and bases in the Pacific. These include around 40 2,900km-range DF-3A (CSS-2) ballistic missiles, each armed with a 3.3Mt warhead; 12 DF-4 (CSS-3) 5,500km-range missiles with 3.3Mt warheads; and some 48 DF-21A (CSS-5), 1,800km-range missiles, each armed with a 200–300kt warhead. The PLA Air Force also has around 100 Hong-6 (B-6) nuclear-capable bombers (with a range of 1,300km) and 30 Qian-5 (A-5A) bombers (with a range of 400km). China is also believed to possess some 120 tactical nuclear weapons deliverable by short-range ballistic missiles (DF-15 [CSS-6] and DF-11 [CSS-7]), artillery or as atomic demolition munitions. In addition to these land-based weapons, China has 12 submarine-launched ballistic missiles, each capable of delivering a single 200–300kt warhead up to 1,000 kilometres, deployed aboard its sole *Xia* nuclear-powered ballistic missile submarine (Norris and Kristensen 2003: 79).

Although the purpose of this arsenal is to provide the nation with a 'minimum deterrent', China's political and military leadership must nonetheless seriously doubt whether the PLA's largely obsolete, predominantly land-based nuclear forces really give them a credible second-strike capability. China's main deterrent against the continental United States, its ageing DF-5 (CSS-4) ICBMs, are liquid-fuelled and deployed in silos (Norris and Kristensen 2003: 77). These missiles are usually maintained unfuelled and unmated to their warheads (Gill and Mulvenon 2000: 41). US analysts estimate that it could take from two to four hours to load fuel into their tanks and install their warheads. Moreover, China does not currently have space-based or land-based early warning systems (ibid.). The net result is that, in the assessment of Bates Gill and James Mulvenon (ibid.), 'in the past, the limited numbers, low level of readiness, and slow response times of China's land-based missiles and bombers left China vulnerable to an overwhelming and incapacitating first strike'. This meant that China's nuclear forces failed to meet one of the basic requirements of deterrence: 'that at least part [of them] must appear to be able to survive an attack and launch one of their own' (Waltz 2003: 20).

Even before the Gulf War provided such a dramatic demonstration of the unprecedented ability of US military forces' space-based intelligence, surveillance and reconnaissance (ISR) assets to locate and identify enemy targets, Chinese strategists were becoming increasingly pessimistic about the ability of the locations of their fixed, land-based nuclear launch sites to evade American detection. Liu Huaqing judged in 1984 that:

> In the face of a large-scale nuclear attack, only less than 10 per cent of the coastal launching silos will survive, whereas submarines armed with ballistic missiles can use the surface of the sea to protect and cover themselves, preserve the nuclear offensive force, and play a deterrent and containment role.
>
> (You 1999: 97)

In anticipation of this increasing vulnerability of China's strategic nuclear forces, however, since the mid-1970s Chinese missile engineers have been working to produce a range of solid-fuelled, mobile missiles which, as the US problems with locating and targeting mobile Scud missiles in the Gulf War revealed, are less vulnerable to pre-emptive first strikes. The first of these is the DF-31, with a range of 8,000 kilometres and a payload of 700 kilograms, which was tested for the first time in August 1999 (Gill and Mulvenon 2000: 48). The follow-on to the DF-31, the DF-31A, is a three-stage, solid-propellant ICBM with a range of 12,000 kilometres, capable of striking targets anywhere in the continental United States (Norris and Kristensen 2003: 77). The DF-31 will be deployed on mobile transporter-erector-launchers (TEL), and it is likely that the DF-31A will be similarly deployed. According to the Pentagon, the deployment of the DF-31 will begin later this decade (USDoD 2003: 37), while the DF-31A will probably replace the DF-5 (CSS-4) ICBMs from 2006 to 2010 (Norris and Kristensen 2003: 77). The CIA reportedly predicts that by 2015, 'most' of China's missile force will be mobile (Norris and Kristensen 2003: 77).

Thus China's political and military leaders have some grounds for supposing that the credibility of China's nuclear deterrent against the United States may be restored once these new mobile strategic systems become operational. This may, however, be only a temporary reprieve until the United States starts to deploy ballistic missile defence systems and advances in technology increase the ability of US military forces to attack a variety of mobile targets, such as SAM radars, tactical ballistic missiles and armoured vehicles (Cote 1999). Contemporary PLA strategic planners are doubtless well aware of the growing confidence of American military experts that with the use of Integrated Global Positioning System (GPS) and Inertial Navigation System (INS) guidance, US military forces will soon be able 'to guide weapons of any range, precisely, night or day, cloudy or clear, to any point on the surface of the earth' – thereby essentially 'solving' the fixed-target problem (ibid.).

Since the poor performance of US military reconnaissance assets to detect mobile Scud missile launchers during the Gulf War, the Pentagon has sponsored the research and development of a space-based system for detecting moving targets on the ground. In 1998, the US Air Force, the National Reconnaissance Office (NRO) and the Defense Advanced Research Projects Agency (DARPA) established a programme, called 'Discoverer II', to develop a constellation of 24 satellites which would use synthetic aperture radar (SAR) and Ground Moving Target Indication (GMTI) collection to obtain very high resolution elevation data and highly accurate radar imagery of most areas of the earth (Federation of American Scientists [FAS] 2000a). The Discoverer II project was killed by Congress in 2000 because of its high cost. But elements of it have survived as a major new funding priority for the US Department of Defense, in partnership with the US intelligence community, in the form of a programme for the development of a space-based radar (SBR) for identifying and tracking mobile targets on the ground and at sea (United States General Accounting Office [GAO] 2004). In his 2005 financial year budget request to Congress, US Secretary of Defense Donald Rumsfeld included SBR, which he described as the only system with the proven capability 'to monitor both fixed and mobile targets, deep behind enemy lines and over denied areas, in any kind of weather', among the Pentagon's critical additional funding requests to strengthen intelligence and persistent surveillance (USDoD 2004a).

Even though the problem of locating, identifying and attacking mobile and moving targets is far from being solved, Chinese strategic planners must still realise that the days of relative invulnerability of ground-based mobile missile launchers are numbered. Moreover, even if the United States' ISR assets are unable to determine the precise location of China's mobile, land-based ICBMs, knowledge of their approximate location may be sufficient to permit their interception soon after launching by boost-phase missile defence capabilities. Chinese strategic planners must therefore consider that it is only a matter of time before the survivability of the PLA's new, land-based mobile strategic nuclear forces once again reduces the credibility of China's ability to deter the United States to unacceptable levels.

As Liu Huaqing implied two decades ago, to preserve the credibility of China's strategic deterrence against the United States Beijing will eventually have no choice but to follow the pattern adopted by the other four major nuclear powers by transforming its nuclear force structure from one based predominantly on its land-based ballistic missiles to a predominantly sea-based submarine-launched ballistic missile (SLBM) force. In fact, the PRC's political and military leadership initiated the development of an SLBM system as long ago as 1958, when Marshal Nie Rongzhen, the leader of weapons research and development, submitted a formal recommendation for commencing a nuclear submarine and SLBM programme to Mao and the central leadership.

At the time, this decision may have reflected a politically motivated desire to emulate the nuclear force structures of other major powers more than any rational assessment of the relative merits of land-based and sea-based nuclear deterrence forces (Lewis and Xue 1994: 5). The United States had begun its Fleet Ballistic Missile System programme in the mid-1950s, and its first SSBN, the *George Washington*, armed with 16 *Polaris* A-1 missiles, entered into service in November 1960 (Miller 1998: 113). The Soviet Union's first sea-based ballistic missile system, an army SS-1 ('Scud') missile converted to naval use aboard a diesel-electric *Zulu*-class submarine, entered into service in 1959 (ibid.: 117). Moscow's first true SSBN, a system combining *Hotel*-class nuclear-powered submarines and SS-N-4 ballistic missiles, first entered into service in 1960 (ibid.: 118). The United Kingdom's first SLBM system entered into service in June 1968, and that of France in January 1972 (ibid.: 405). It took three decades – from its commencement in 1958, the launching of the *Xia* SSBN at Huludao Naval Base in 1981, and the first successful launching of the Julang I (JL-I) SLBM from the *Xia* in September 1988 – for China to produce its first working SLBM system. But even now, after almost half a century of development, China is the only one of the five major nuclear powers which has yet to develop a truly effective sea-based nuclear deterrent capability – at least as far as its primary potential nuclear adversary is concerned.

At most, the sea-based leg of China's current nuclear triad would have only limited deterrent effect against the United States. Its single Type 0–92, *Xia*-class SSBN with its 12 CSS-N-3 (JL-1) intermediate range SLBMs (1,800km) has only conducted one live missile firing, and is thought not to have left port since then because of problems with its nuclear reactor and propulsion systems. According to Norris and Kristensen (2003: 78) the *Xia* is thought not to be fully operational, even after a recent four-year overhaul, and it has never sailed beyond China's regional waters. This first generation SSBN is, as Shambaugh (2002: 271) describes it, 'compared to the ultraquiet *Ohio*-class SSBNs, a relic – no match for Japanese SSNs or SSKs'. In an emergency, the *Xia* could launch its missiles without putting to sea, and could plausibly pose a threat to US bases in Japan or South Korea – although even this threat would be mitigated by the deployment of a reliable theatre missile defence system. But high-value targets on the territory of the United States, including military bases on Guam, would be well beyond their range. As Gill and Mulvenon (2000: 38) observe of China's current sea-based nuclear deterrence system:

> The limited range of the [Jl-1] missile, the problems it has had in deployment and operation, and the limited experience of the Chinese in long-range submarine operations limits the value of this system as a strategic weapon. Beijing may also have learned some valuable negative lessons from the experience of the Soviet Union, whose SSBN force was forced to retreat to bastions by a superior US attack submarine fleet.

The range of its SLBMs and the vulnerability to detection and interdiction of its SSBNs will become increasingly important considerations for Chinese strategists as China advances towards the realisation of its ambition to develop a more credible second-strike strategic nuclear deterrent capability. Most Chinese scholars believe that the most effective way of giving China's nuclear deterrent greater invulnerability is to build more sea-based missiles (Tomkins 2003). The imperative for China to place greater reliance on the maritime component of its nuclear deterrence forces will also increase with the deployment by the United States of effective missile defence systems to protect the US homeland and US bases and allies in East Asia. Submarine-launched missiles pose a harder problem for any anti-missile system because of their short flight times and uncertain launching locations (Hezlet 1967: 250). So the coming decades are likely to see a shift in emphasis of the PLA's efforts to modernise its strategic nuclear forces from its new land-based, road-mobile DF-31 intermediate-range ballistic missiles and future DF-31A ICBMs towards a more effective SSBN force equipped with longer-range SLBMs. This shift in emphasis is perhaps already evident in the amount of space devoted to the discussion of SSBN operations in China's journal of naval warfare, *Jianchuan Zhishi* (Goldstein and Murray 2004: 172).

But geography has dealt the PRC a weak hand when it comes to operating submarine-based systems as an effective strategic deterrent against the United States. Unlike the other nuclear powers whose strategic nuclear deterrence capabilities are invested partly (in the case of the United States, Russia and France) or wholly (in the case the United Kingdom) in their SLBM systems, the PRC's SSBNs cannot reach the open-ocean patrol areas without passing through the choke points formed by the chain of islands – from the Kuriles, through the Japanese home islands and the Ryukyus, to Taiwan and the Philippine archipelago – which enclose China's littoral seas. The control of these islands by members of the network of US Asia-Pacific security partners enables the United States to establish a distant blockade of the Chinese fleet to prevent the projection of Chinese sea power beyond its continental shelf, in much the same way that Great Britain established a distant blockade across the exits from the North Sea to prevent the German High Seas Fleet from gaining access to the Atlantic during the First World War, and that NATO established a barrier across the Greenland–Iceland–UK gap to prevent the entry of the Soviet fleet into the Atlantic during the Cold War. Although geography also endows Russia (and the Soviet Union) with relatively restricted access to the open ocean for its SSBNs, its Pacific Fleet SSBNs, based at Ryabachi near Petropavlosk on the Kamchatka Peninsula, have unrestricted access to the open ocean, while its Northern Fleet SSBNs, based on the Kola Peninsula, can transit the Barents Sea to find refuge under the ice of the Arctic Ocean. In fact, once the range of their SLBMs exceeded 8,000 kilometres (with the Delta-I/SS-N-8 combination which entered into service in the early 1970s), Soviet/Russian SSBNs no longer needed to leave the semi-enclosed Sea of Okhotsk in the east and the Barents Sea in the west, where they were relatively invulnerable to Western ASW activities. From these 'bastions', Soviet SSBNs could patrol under the protection of naval and land-based air forces, while their 8,000km-range SLBMs enabled them to strike targets throughout the continental United States (Miller 1998: 122).

China has begun work on a new SSBN called Project 094 (Norris and Kristensen 2003: 78). When the PLA Navy eventually deploys its second generation Type-094 SSBNs they will be armed with 16 JL-2 ballistic missiles with a range of 8,000 kilometres (Shambaugh 2002: 72). But to pose a credible deterrent threat to the continental United States, these Type-094 SSBNs would still have to deploy to deep-ocean patrol areas in the Pacific to launch their missiles. A credible threat to target, say, Los Angeles, would require an ability to launch JL-2 missiles from an area east of the Kurile Islands in the north Pacific or from east of the second island chain in the central Pacific. Launched from the China Seas within the confines of the first island chain, a JL-2 missile could at best pose a theatre-level threat, reaching targets in US territory in Guam, Hawaii, Alaska and perhaps the extreme northwest coast of the continental United States. The PLA Navy would need an SLBM of intercontinental range, such as the 12,000km-range Trident II, to be able to threaten strategic targets in the continental United States from launch areas in China's littoral seas.[3] Even then, however, as long as the first island chain is controlled by the forces opposed to the PRC and bound by security ties to the United States, or until the PLA Navy is able to establish unchallenged control over its littoral waters and surrounding airspace, Chinese SSBNs will be potentially vulnerable to hostile ASW activities and will not be able to enjoy the kind of sanctuary with which geography blessed Soviet/Russian SSBNs in their Barents and Okhotsk bastions.

So for the foreseeable future, China's ability to deploy a credible nuclear deterrent against the United States will depend increasingly on the ability of its SSBNs to pass undetected across the trip wires and ASW barriers across their transit routes through the choke points between the islands enclosing the China seas. During the first decades of the Cold War NATO positioned attack submarines in the strategic straits through which Soviet submarines passed on their way to their patrol areas (Friedman 2001: 87). Chinese strategic planners would have to assume the presence in these gaps of arrays of fixed sub-surface, and mobile surface and airborne submarine detection systems, such as the Sound Surveillance System (SOSUS) used during the Cold War and its modern equivalent, the Integrated Undersea Surveillance System (IUSS), as well as anti-submarine mines such as CAPTOR (encapsulated torpedo), similar to those employed by the United States and its allies during the Cold War to detect the passage of Soviet SSBNs through coastal waters and choke points on their way to their patrol areas in the Atlantic and Pacific Oceans.

US Navy ASW doctrine anticipating requirements for ASW in the period from 2010 to 2015 envisages the deployment along anti-submarine barriers of a new range of smart ASW mines capable of distinguishing friend from foe over distances of up to two nautical miles (US Naval Doctrine Command 1998). If the United States obtains Japanese permission for US military access to the 10,000-foot concrete runway and to build a port on the island of Shimoji Shima halfway between Okinawa and Taiwan, these facilities could be used to reinforce ASW barriers linking the first chain of islands enclosing the China seas, particularly as a base for long-range maritime patrol aircraft (Brooke 2004).

If China's new Type-094 SSBNs incorporate quieting technologies similar to those of the new Type-093 SSNs, which the Pentagon considers to be comparable with the Soviet second generation *Victor III* SSN, they could pose a significant challenge to the ASW technologies and tactics honed by the United States to counter second generation Soviet nuclear submarines during the 1970s and early 1980s. The quieting technologies first incorporated into the *Victor III*, first deployed in 1978, enabled it to elude SOSUS and frustrate efforts by tactical ASW platforms using passive sonar to establish and maintain contact with it (Cote 2003: 69). Despite the significant advances in stealth likely to be demonstrated by China's second generation SSBNs, geography and new ASW technology such as the Advanced Deployable System (ADS), a passive ocean-bottom array linked to shore-based data processing facilities by fibre-optic cable, and Distant Thunder, a powerful signal processing system, may still enable the United States Navy to create effective ASW barriers across the choke points between the first island chain to prevent Chinese nuclear submarines from deploying into open ocean areas of the Pacific during war.

Establishing fixed strategic defences across the passages between the islands along the Chinese continental shelf would present less of a challenge to the United States Navy than the establishment of similar defences across the passages between Greenland, Iceland, the Faeroes and the United Kingdom, which were in total about 600 miles (966 kilometres) wide (Moore and Compton-Hall 1987: 159). The political and physical geography of the western Pacific is well suited to the establishment of ASW barriers to restrict the egress of continental navies to the open ocean. The SOSUS arrays established near the Kurile Islands and elsewhere in the western Pacific during the Cold War were considered the most threatening to Soviet SSBNs (Stefanick 1987: 39). And in 1943, the Japanese generally succeeded in keeping American submarines out of their semi-enclosed seas by establishing mine barriers from the northern tip of Taiwan to the Japanese home islands (Hezlet 1967: 216). Soviet sources estimated that the establishment the Greenland–Iceland–UK barrier would require four long-range maritime patrol aircraft to be constantly on station over the gap, co-operating with seven patrolling submarines, and the positioning of 1,000 CAPTOR mines between Greenland and Iceland. The Soviets calculated that this would ensure an attrition rate of 30 per cent against noisy submarines (Moore and Compton-Hall 1987: 159). The resources and technical requirements for the creation of ASW barriers across the gaps in the East Asian first island chain would be well within the means of US and allied military forces. Moreover, the proximity of naval and air bases on Okinawa and in the Japanese home islands would facilitate the defence of the fixed and mobile ASW surveillance systems against attempts to neutralise them.

The obstacles posed by geography to the PRC's future ability to develop an effective strategic nuclear deterrent against the United States would be very much reduced if Beijing were to obtain control of the island of Taiwan. Taiwan forms a key link in the island chain which follows the edge of the East Asian continental shelf and bars direct access by China to the open ocean. On the other hand, Taiwan also has a number of ports on its east coast, including the major port of Chi-lung (Keelung), home to 15 per cent of the ROC's naval forces, and Su-ao and Hua-lien, homeports respectively to 8 per cent and 2 per cent of Taiwanese naval units (Cabestan 2003: 434). These ports give access directly to the open ocean beyond the continental shelf and the chain of islands which enclose the Yellow, East and South China Seas. Moreover, to strengthen its ability to mount a counter-attack against the mainland, the ROC Navy has begun the construction of a new base at the east coast port of T'ai-tung by hollowing out the mountainside in order to build a secure base for the submarines that Taipei hopes to acquire (Cabestan 2003: 112). Should mainland Chinese forces succeed in gaining control of the island of Taiwan, they would possess an invaluable platform for the projection of Chinese offensive sea power beyond its littoral seas into the wider region of maritime East Asia and the Pacific.

Just as control of the Atlantic ports of Norway and western France was an essential for Germany to mount an effective challenge for naval supremacy in the Atlantic Ocean during the Second World War, control of Taiwan – the only part of greater China which has direct access to the open ocean – would be a crucial step towards the realisation of any future

Chinese ambition to challenge the US Navy's command of the western Pacific. The development of such a capability would be a fundamental condition for the realisation of any Chinese ambition to supplant the United States as the dominant power in East Asia.

As long as the island of Taiwan is controlled by forces opposed to the regime in power in mainland China, it will constitute an obstacle to the consolidation of Chinese strategic power in East Asia and any attempt by Beijing to challenge the United States' current regional primacy. For the United States whose avowed strategic interest is – in the words of the 2001 Quadrennial Defense Review – to preclude 'the hostile domination of the East Asian littoral' (defined as the region stretching from south of Japan through Australia into the Bay of Bengal) (USDoD 2001: 2), the strategic geography of East Asia provides an advantage similar to that enjoyed by the United States and its maritime alliance over the Soviet Union during the Cold War. This advantage was described by Paul H. Nitze (1998), then US Secretary of the Navy, in his 1964 graduation address to the Naval War College. Observing that a major problem for his generation of naval commanders was the lack of progress in anti-submarine warfare, and pessimistic about the prospects for the generations of the 1970s and 1980s to 'reduce to manageable proportions the problem of the opacity of the seas', he nevertheless noted one very remarkable advantage:

> the fact that the Atlantic littoral is composed of entirely free nations. Any potential aggressor must send his submarines through straits and passages that are susceptible to mining and patrolling. We can, therefore, expect that this geographical advantage would exact very great attrition upon enemy submarines in a limited war at sea.

From the other side of the East–West divide, during a conversation with Mao Zedong in July 1958, Khrushchev evoked the constraints imposed by geography on Soviet submarine operations. The main task for a submarine fleet, as Khrushchev saw it, was not to attack the enemy's surface fleet, but to attack his ports and industrial centres. However, it was not easy for boats based in Murmansk to reach America and they were liable to be intercepted by Britain and Iceland. 'Vladivostok was better,' according to Khrushchev, 'but there as well we are squeezed by Sakhalin and the Kurile Islands – they defend us, but also allow the enemy's submarines to monitor the exit of our submarines.' Following Mao's vigorous rebuff of Moscow's proposal for a joint Sino-Soviet submarine flotilla, Khrushchev sought Mao's agreement to give Soviet submarines basing facilities on what he described as China's 'vast coastline and access to open seas, from where it would be easy to conduct the submarine war with America' (Woodrow Wilson International Center 1958).

Apart from their obvious climatic advantage over the USSR's periodically ice-bound Pacific naval base at Vladivostok, however, Chinese ports would not have enabled Soviet submarines to reach the open waters of the Pacific Ocean any more easily than from Vladivostok. With its shores washed by the pelagic China seas, semi-enclosed by the chain of islands stretching from Japan, the Ryukyus, Taiwan and the Philippine archipelago, all of which – unlike Sakhalin and the Kuriles in the Soviet case – are under the control of potential Chinese adversaries whose security depends on the American network of Pacific defence ties, the freedom of action of the PRC's naval forces is even more circumscribed than that of the Soviet Union during the Cold War. However, this task of containing Chinese SSBNs would become much more difficult if the PLA Navy were to acquire bases in Taiwan, situated as it is on the edge of the continental shelf, with ports on its east coast enjoying unrestricted access to the open ocean. From this perspective, Beijing's essential strategic interest in being able to assert effective sovereignty over Taiwan is to achieve that classic goal of naval strategy – command of the sea. And command of the sea, as Corbett (1911: 94) emphatically pointed out, 'means nothing but the control of maritime communications, whether for commercial or military purposes'. In this case, control of Taiwan would immeasurably assist the PLA Navy to assert sufficient control over the transit routes of its SSBNs to enable them to reach their patrol areas in the central and eastern Pacific from where they would be able to bring their strategic offensive power to bear directly on the United States homeland.

For the foreseeable future, however, technology is likely to prevent China from following the Soviet example of developing SLBMs of such range that they could be launched from SSBNs patrolling within local waters. This means that the only option available to Beijing to shore up the credibility of the PRC's strategic nuclear deterrent against the United States is to try to improve the poor hand that it has been dealt by geography. Bringing the island of Taiwan back within the sovereignty of the Chinese mainland, by persuasion, coercion or by main force, is therefore an essential to any attempt by China to supplant the United States as the predominant power in East Asia. Thus as long as Taiwan remains beyond Beijing's control, any attempt by the PLA Navy to project force beyond its littoral waters – which would require it to secure a working control of the sea at least out to the open-ocean patrol areas of its SSBNs – will be severely handicapped by its restricted access to the open Pacific Ocean beyond the first island chain. At the strategic level, therefore, geography leaves Beijing little choice but to adopt a defensive naval posture beyond the narrow seas off the Chinese coast.

3

CHINA'S NEW MARITIME
STRATEGY

Liu Huaqing's strategic reformation

As a consequence of its new strategic and economic circumstances, from 1985 Beijing began a radical reform of its military forces. Ground forces were reduced in number. The latest round of cuts, announced in September 2003 by CMC Chairman Jiang Zemin, will reduce PLA personnel numbers by 200,000 before 2005, following an earlier reduction of 500,000 between 1996 and 2000. According to the PLA, this will reduce the PRC's army to 2.3 million soldiers (*People's Daily*, 2003). With the resources liberated by this measure, the PLA introduced recruitment and training programmes to increase the level of technical proficiency of its personnel and to equip its forces with modern weapons. Given the traditional technology-intensive nature of naval forces, the PLA Navy will be one of the principal beneficiaries of these reforms.

As Commander-in-Chief of the PLA Navy from 1982 until 1988, Liu Huaqing received instructions from the CMC to prepare the navy for two high-probability missions in the near-to-medium term – the safeguarding of China's territorial integrity and the conduct of a possible blockade against Taiwan – as well as for two longer-term missions – the prevention of a large-scale invasion and the deterrence of nuclear attack (Lewis and Xue 1994: 226).

Liu identified two maritime zones which the Chinese Navy should be capable of controlling. The first zone, the control of which represented the first phase of Liu's strategy, encompasses the Yellow Sea opposite Japan and the Korean Peninsula; the western part of the East China Sea, including Taiwan; and the South China Sea. China's vital national interests are at stake in these geographic areas: its territorial claims, its maritime natural resources and its coastal defence. Chinese strategists describe this zone as delimited by 'the first island chain', a north–south line which passes through the Aleutian Islands, the Kurile Islands, the Japanese archipelago, the Ryukyu Islands, Taiwan, the Philippines and Indonesia. In his revision of China's maritime strategy, Liu determined that the PLA Navy should aim to be capable of controlling this zone by the year 2000, the end of the first phase of the implementation of the new strategy. In the 1980s, Liu Huaqing indicated that the area to be 'safeguarded' by the PLA Navy extended out to 200 nautical miles from the coast. He reportedly increased this distance later to 600 nautical miles (Kondapalli 2001: 3).

The second maritime zone, and the second phase of Liu's maritime strategy, is delineated by the 'second island chain'. This is a north–south line which goes through the Kurile Islands and Japan, and then takes a more eastern course through the Bonin, Mariana and Caroline Islands. The control of this geographic zone, which Liu determined the PLA Navy should aim to achieve by 2020, would secure for the PRC control of the whole of East Asia's oceanic area. The achievement of this goal implicitly assumes the withdrawal of the United States' military presence from the region.

The third phase of Liu Huaqing's maritime strategy was to create a blue-water navy capable of exercising a global influence by 2050.

A maritime strategy conceptualised as a series of concentric defence perimeters – with the navy constituting, in Jiang Zemin's words, a 'Great Wall at sea' – betrays the enduring influence of China's continental strategic culture on China's political and military policy-makers. Since at least the fourth century BC, when the Xiong-nu nomadic tribes on China's north-eastern borders started raiding China's settled farmlands, the need to defend the interior Chinese heartland against threats coming from an alien periphery has provided the basic conceptual framework for structuring Chinese approaches to national strategic and foreign policy. It has created an indelible mark on Chinese strategic thinking which, in Fairbank's words (1974: 12), 'at times might even become a Great Wall mentality far more deep-lying than the temporary aberration of General Maginot'.

As Michael Swaine and Ashley Tellis (2000: 24) put it, 'historically, the defence of [the] Chinese heartland required efforts by the Chinese state to directly control, influence or neutralise a very large periphery surrounding it'. The concept of a maritime strategy based on a series of concentric defence zones emerges clearly in comments made in 1988 by Zhang Lianzhong, Liu Huaqing's successor as Commander-in-Chief of the PLA Navy:

> [First,] the exterior perimeter [encompasses] the seas out to the first chain of islands. This region will be defended by conventional and nuclear submarines (some of which will be armed with anti-ship missiles), by naval medium-range aircraft and by surface warships. The submarines will play a dynamic role to ensure defence in depth, including the laying of mines in the enemy's sea lines of communi-cation. The middle defence perimeter extends 150 miles from the coast and comes within, but in most cases does not reach, the first chain of islands. Anti-ship aircraft, destroyers and escort vessels will carry the main burden in this area. The interior defence perimeter extends to sixty miles from the coast. This will be the

theatre of operations for the main naval Air Force, fast-attack boats and land-based anti-ship missile units.

(Cole 2001: 167–8)

As J. C. Wylie (1967: 49) has observed, the strategic conceptions of the soldier, the sailor and the airman are shaped by the environments in which they operate: 'where the sailor and the airman think in terms of an entire world, the soldier at work thinks in terms of theaters, in terms of campaigns, or in terms of battles'. The soldier's approach to strategic planning and operations is dominated by the limitations of geography and terrain. Land warfare strategy is conceived in terms of distinct, geographically limited theatres, whereas for air and maritime strategy such geographically defined limits are somewhat more arbitrary. These differences in conceptual approaches to strategy are naturally likely to be more pronounced in those strategic cultures, like those of the United States and Great Britain, where there is a long-established tradition of intercontinental, transoceanic operations. It is not surprising, therefore, that the conceptual basis of China's strategic planning is heavily impregnated with a continental, land warfare approach.

Liu's new maritime strategy was in fact more an extension seawards of the doctrinal thinking initiated by Deng Xiaoping in 1975 than a radical innovation in Chinese strategic thinking. On his return to power in 1977, Deng began a revision of Mao Zedong's 'people's war' doctrine which had guided the PLA's strategic posture and operational concepts since Mao won control of the CCP's military forces after the Long March in 1935. Deng questioned the contemporary relevance of Mao's fundamental precept of 'luring the enemy in deep' and defeating them in a war of attrition, and proposed instead that the PLA should adopt a doctrine of 'active defence' of China's frontiers, rather than the Maoist one of defence-in-depth (Shambaugh 2002: 62). Following the PLA's manifestly poor performance during the war against Vietnam in 1979, its operative doctrine became one of 'people's war under modern conditions', which emphasised 'active' frontier defence with modern weapons, rather than in-depth defence relying on superior numbers. 'Active defence' underlined the utility of offensive operations, seizing the strategic initiative and pre-emptive attacks (Shambaugh 2002: 65). With the fading of the Cold War superpower confrontation, this doctrine was further revised in 1985 to become one of 'limited war', based on the assessment that the most likely conflicts for which the PLA should be prepared would be geographically limited wars in China's region usually involving only two combatants (Shambaugh 2002: 64). The Gulf War was the catalyst for yet another revision of the PLA's basic doctrine in 1991 when China's political and military leaders were struck with the evidence of the revolution in military affairs which enabled American forces to defeat Iraqi forces so swiftly and decisively. As a result, the PLA adopted a doctrine of 'limited war under high technology conditions' designed to enable its forces to fight local, limited wars against an enemy employing electronics, computers, satellites and stealth to deliver massive, high-precision, stand-off firepower.

But even while Chinese naval strategy is still fundamentally shaped by the continentalist framework from which it springs, as well as by Chinese strategists' more recent analyses of the United States' essentially land-warfare campaigns since the end of the Cold War, Liu's achievement is to have begun the work of breaking China's centuries-old mould of land-bound strategic thinking and to have launched the PLA Navy conceptually on a new path towards an open-ocean navy. In addition to expanding the geographic scope of the PLA Navy's activities from green water to blue water,[1] Liu's new strategy also shifted its posture from one of reactive defence to one which gives greater emphasis to operational and tactical offensive action. The guiding principle of the new maritime strategy was to transform the main mission of the PLA Navy from one of coastal defence to one of 'offshore active defence'. The principal elements of this strategy are 'a stubborn defensive posture near the shore, mobile warfare at sea, and surprise guerrilla-like tactics at sea' (Cole 2001: 166). The aim of China's maritime strategy was no longer to lure the enemy deep into Chinese territorial waters to engage it in a 'people's guerrilla war', but to confront the enemy in the outer approaches and stop its advance well before it reached coastal waters (You 1999: 166).

Liu Huaqing's vision of a blue-water navy for China was developed at a time when the question of the future of Taiwan appeared less urgent and critical to the PRC's leadership than it is today. In the 1980s and early 1990s, Beijing's more urgent sovereignty concerns were focused on the South China Sea. But in 1995, following Taiwanese president Lee Teng-hui's visit to the United States, Jiang Zemin announced that the 'current situation has placed new demands on building the navy' and gave the PLA Navy 'three major directions':

- place naval building in an important position and accelerate the pace of naval modernisation;
- ensure the security of China's coastal defence;
- promote the accomplishment of the great cause of reunification of the motherland.

(Kondapalli 2001: 5)

The re-emergence of the Taiwan issue as Beijing's principal strategic concern in the mid-1990s deflected the PLA Navy from the course charted for it by Liu for the achievement of a capability to exercise control of the China seas out to the first island chain by 2000 and the second island chain by 2020. The new priority accorded to Taiwan, particularly since the 1995–6 crisis, prompted a switch in the PLA Navy's general and more positive aim of acquiring the capabilities to enable it to exercise control of the seas out to the first island chain, to the less ambitious and more negative aim of being able effectively to deny the control of these seas to hostile forces. This shift in goals from sea control to sea denial is reflected in the apparent loss of momentum in China's interest in acquiring an aircraft carrier capability, long-range land-based bombers and transport aircraft, and its reinvigorated interest in acquiring the instruments of sea denial in the form of advanced weapons platforms such as the *Sovremenny* destroyers, *Kilo*-class submarines and a combat aircraft airborne-refuelling capability. Aircraft carriers could be useful to Beijing's efforts to assert its sovereignty over the distant reefs and rocky outcrops of the South China Sea, but would have little advantage over land-based aircraft in conflict in the narrow seas surrounding Taiwan (Storey and You 2004).

If, as seems increasingly likely, Beijing has put plans to acquire an aircraft carrier on hold, it may be because Chinese strategic planners have adopted a similar logic to that of the Soviet Navy when, soon after Gorshkov succeeded Admiral N. G. Kuznetsov as Commander-in-Chief in 1955, Moscow apparently decided to forgo the building of large attack carriers, cancelling the carrier construction project that Stalin had revived following World War II (Wolfe 1973: 254). At the time, Gorshkov, echoing similar well-publicised remarks made by Khrushchev, justified this decision on the increasing vulnerability of large carriers in the nuclear-missile age and the assertion that carriers could not 'compare with the striking power of under-water and air forces' (Wolfe 1973: 254). Western analysts of the Soviet Navy attributed this reversal of policy to the growing strategic threat from US strike carriers at a time when the Soviet Union's resources for countering it by building up carrier forces of its own remained limited (Wolfe 1973: 251). During the Khrushchev years, before the Polaris SLBMs superseded US aircraft carrier strike groups as the primary threat, the principal Soviet naval effort was directed towards the building of submarines (Wolfe 1973: 255). One of the first manifestations of this new focus on countering the carrier threat was the modification of *Whiskey*-class long-range attack submarines to carry anti-ship cruise missiles (Wolfe 1973: 256). Beijing's decision to purchase its first batch of *Kilo*-class submarines shortly after the shock of the deployment of the US carrier groups to the Taiwan Strait during the 1996 crisis seems to have coincided with the shelving – for the time being at least – of plans to develop an aircraft carrier capability.

In practical terms, therefore, at least in the short term, it seems probable that the primary focus of the PLA Navy's development efforts is less to enable it to project power outside China's immediate region than to strengthen its ability to dominate its immediate vicinity and deny access to any hostile powers to an area within some 200 nautical miles of the Chinese coast (Lexington Institute 2004: 13).

The strategic missions of the PLA Navy

Thus the PLA Navy's fundamental mission remains the defence of China and its maritime frontier. This is a difficult task by virtue of the fact that the coast of China extends from the Korean Peninsula in the north to the Gulf of Tonkin in the south. It is made even more difficult as a result of the strong growth of maritime activity in the PLA Navy's areas of operation since the opening of the Chinese economy in the 1980s. The PLA Navy's regular peacetime missions are those that relate to coastal surveillance – controlling piracy and narcotics and arms trafficking, fisheries protection and the protection of maritime communications. For a state like China, which nurtures ambitions to recover its former status as the pre-eminent regional power, naval diplomacy is also an important part of the PLA Navy's functions.

For times of war, Liu Huaqing set the PLA Navy the goal of attaining the capabilities to undertake four basic missions:

- securing sea control in the major battle directions in China's offshore waters;
- blockading major sea lines of communication effectively within a required span of time, in waters encompassing China's maritime territories;
- initiating major sea battles in waters adjacent to China's maritime territories; and
- waging reliable nuclear retaliatory strikes.

(You 1999: 168)

To perform the last of these four basic missions, the PLA Navy currently relies on its strategic submarines: the single Type 0–92 *Xia*-class SSBN, and a conventionally powered Soviet *Golf*-class submarine capable of launching a JL-1 intermediate-range ballistic missile. This role will be performed in the future by the new Type 0–94 SSBN which will be capable of carrying 16 JL-2 type ballistic missiles with a range of 8,000 kilometres and a 250kt warhead.

With respect to the first three missions, those to be performed by China's conventional naval forces, the Chinese officers' naval manual translates Liu's broad instructions into more concrete operational language and creates the additional task of assisting the conduct of an amphibious assault:

- wipe out enemy naval vessels;
- participate in anti-submarine warfare;
- transport and guarantee the [mission of the] landing troops on enemy's shores;
- foil the enemy's objective of attacking one's coasts;
- reconnoitre the seas, patrol, be on alert against landmines, lay and clear mines, prepare for convoy duties, rescue of civilians on seas, transport men and material.

(Kondapalli 2001: 34)

The PLA Navy's conventional tactical submarine fleet has a major role in carrying out the first three of the strategic missions established by Liu. Thus the officers' naval manual also contains specific instructions for the submarine corps. In addition to the nuclear deterrence mission, the manual sets out the responsibilities of the submarine corps as:

- wipe out enemy transport vessels and large and medium-type attack craft;
- damage, destroy enemy naval bases, harbours and (strive to fulfil) the objective of attacking the (enemy) coasts;
- carry out reconnaissance on seas, do patrolling, be on alert against mines, lay and clear mines, (prepare for) convoy, rescue and transportation of personnel and materials, etc.

(ibid.: 48)

Not surprisingly, there is little information available from open sources about the clandestine reconnaissance missions of the Chinese submarine fleet, but Chinese submarines undoubtedly play an important role in intelligence, surveillance and reconnaissance. China is known to have one of Asia's most extensive signals intelligence capabilities (Tripplet 2000: 83). The PLA Navy operates at least 10 electronic intelligence ships and the PLA Air Force has an estimated 290 reconnaissance/electronic intelligence aircraft (IISS 2003: 155), including a number of EY-8 turboprop reconnaissance aircraft (Shambaugh 2002: 80).

These operational tasks set out in the Chinese naval officers' manual are, naturally enough for such a document, general and abstract. But in terms of specific strategic missions, there are two which the PLA Navy has to be ready to execute at a moment's notice. It is these missions which have a dominant influence on the planning of the PLA Navy's force structure, doctrine, organisation and training. The first of these is the defence of Chinese interests and sovereign territory in the East and South China Seas; the second is the use of force to secure Beijing's political interests with respect to Taiwan.

Taiwan

Since the evaporation of the threat from the Soviet Union and the democratisation of Taiwan, the main aim of China's efforts to modernise its armed forces has been to develop the capabilities that the PLA would require to ensure that force could be used successfully to reassert Beijing's effective sovereignty over the island (Cabestan 2003: 26). This reorientation of China's strategic priorities to place the resolution of the Taiwan issue as the primary objective of the PRC's national strategy was reportedly decided at a meeting of the CMC in May 1993 (Cabestan 2003: 27), although the development of a capacity to conquer Taiwan by force did not become evident until the aftermath of the March 1996 missile crisis (Pillsbury 2001: 6). The importance of this mission for the armed forces of the People's Republic was evident at the 16th Congress of the CCP in November 2002 when, of the group of generals promoted to the CMC, only one (Xu Caihou, Director of the PLA's General Political Department) was not a specialist in Taiwan-related warfare (Cabestan 2003: 29).

Beijing's policy towards Taiwan requires that the PRC is able to present a permanent and credible threat to use military force to prevent the Taiwanese from declaring their independence from China. The challenge for China's political and military leadership is, in John Landry's (2001: 86) words, 'to find the means to deter Taiwan's bid for independence in the short term while developing longer-term capabilities both to seize the island if necessary and hold off external intervention'. Geography dictates that naval forces are necessarily an essential component of this threat. One of the PLA Navy's most important tasks, therefore, is to draw up plans for the use of naval forces against Taiwan.

According to Shambaugh (2002: 325), PLA planning for the use of force against Taiwan focuses on three possibilities: an amphibious assault, a blockade and strategic strikes. While the last of these operational contingencies would give a prominent role to the Second Artillery, with its short-range ballistic missile batteries based in the Nanjing Military Region opposite Taiwan, the PLA Navy would have a central role in the former two contingencies.

Although the PLA exercises amphibious assault scenarios with increasing frequency, its current capabilities preclude this option for the foreseeable future (Shambaugh 2002: 325). Most analysts consider that an amphibious assault of significant size across the Taiwan Strait would be exceptionally difficult, even if China succeeded in achieving air and sea superiority in the Strait, something that is far from certain with the current balance of capabilities between China and Taiwan (Shlapak *et al.* 1999). The meteorological conditions and the currents in the Strait are notoriously difficult. The western coast of the island of Taiwan is made up largely of tidal mudflats which would be ill suited to an amphibious landing, while the eastern coast is defended by steep cliffs. Leaving aside these geographic obstacles, the PLA would have to overcome its almost complete lack of suitable maritime and air transport capabilities before it could seriously consider an amphibious assault of Taiwan. According to Shambaugh (2002: 325):

> The conventional wisdom is that in such landings, a 5:1 numerical advantage is needed (irrespective of terrain): thus the PLA would have to land approximately 1.25 million troops on Taiwan within the first few days of the invasion. This is, of course, impossible. At present, it is believed that the PLAN only has sealift capability to transport one or two divisions and about 300 tanks at a time, far short of the numbers necessary to establish a beachhead on the

heavily fortified western approaches of the island. It would take approximately 600 landing craft nearly two weeks to transport twenty infantry divisions to Taiwan.

The fact that China does not have an amphibious assault capability, and shows no sign of building one, either in its merchant fleet or in the PLA Navy (Cole 2003a: 136), is further evidence to indicate that for the time being at least, Beijing does not consider an amphibious invasion of Taiwan to be a serious possibility.

Other than a campaign of ballistic missile strikes against key military and economic targets, a naval blockade is therefore the most likely means for Beijing to use military force to achieve its strategic ends with respect to Taiwan. In February 2000, the Australian press reported a leaked US intelligence assessment that China intended to blockade the port of Kaohsiung (FAS 2000b). Taiwanese strategists also assess that such an attempt is highly likely. In 1998, in its assessment of the most likely options for the use of force by the PLA, the Ministry of Defence in Taipei included the declaration of a partial naval and air blockade of Taiwan's offshore islands, and the imposition of a comprehensive naval and air blockade in the waters and air space around Taiwan proper (Dreyer 1999). Indeed, Lewis and Xue (1994: 228) note that 'Chinese leaders ... estimate that a blockade of the island would constitute a minimal reaction to any move toward Taiwanese independence'.

Beijing has been contemplating the possibility of blockading Taiwan for at least the past two decades. Lewis and Xue (ibid.) report that in 1984, Deng Xiaoping told US Defense Secretary Caspar Weinberger: 'China does not now have the military forces to invade or occupy Taiwan, but we have the military power to blockade' the Taiwan Strait. Lewis and Xue (ibid.) also note that:

> In the following year, when answering questions on a possible blockade of Taiwan, Hu Yaobang, then head of the Chinese Communist Party, rated the feasibility of such a blockade as low and concluded that China would not have sufficient forces to accomplish it for at least another eight years. But he went on to say: 'Once we decide to enforce the blockade of Taiwan, we do not believe it will be difficult to deal with Taiwan itself,' even though China would have to take into account the possibility of armed intervention of foreign countries.

In its 1999 annual report to Congress, the Pentagon outlined several options or courses of action available to Beijing in the event that it decided to use force against Taiwan – including an interdiction of Taiwan's sea lines of communication (SLOC), and a blockade of Taiwan's ports, a large-scale missile attack and an all-out invasion. With respect to the blockade action, the Pentagon (USDoD 1999) considered that:

> Beijing probably would choose successively more stringent quarantine-blockade actions, beginning with declaring maritime exercise closure areas and stopping Taiwan-flagged merchant vessels operating in the Taiwan Strait. Operations likely would include mine laying and deploying submarines and surface ships to enforce the blockade. Barring third party intervention, the PLAN's quantitative advantage over Taiwan's Navy in surface and sub-surface assets would probably prove overwhelming over time. Taiwan's military forces probably would not be able to keep the island's key ports and SLOCs open in the face of concerted Chinese military action. Taiwan's small surface fleet and four submarines are numerically insufficient to counter China's major surface combatant force and its ASW assets likely would have difficulty defeating a blockade supported by China's large submarine force.

That a blockade rather than an amphibious invasion should be among Beijing's preferred strategies to impose its will on the leadership and people of Taiwan simply reflects the vulnerability of insular polities to this kind of action and their relative invulnerability to physical invasion. This lesson of history was amply demonstrated by Great Britain in the First World War and again, together with Japan and Malta, in the Second World War.

Taiwan is vulnerable to a blockade because it has few natural resources, a smaller economy than China and an almost complete economic dependence on maritime and air commerce (O'Hanlon 2000: 75). The small size of the island, together with the fact that it only has two major ports, one at each end, restricts shipping to a limited number of predictable routes. Using mines and submarines, it would be a relatively easy task for the PLA Navy to prevent movement in and out of Taiwan's main ports of Keelung and Kaohsiung. Chinese submarines normally carry between two and three dozen mines. As Michael O'Hanlon observes, if half the Chinese submarine fleet were able to sow mines around Taiwan without getting themselves sunk, China could deploy as many mines as Iraq did during the first Gulf War (O'Hanlon 2000: 78). And Beijing could justify this action by claiming that the seas around Taiwan formed part of its territorial waters and that mining them was therefore not contrary to international law. Goldstein and Murray (2004: 180) report on a recent article in a Chinese military journal reflecting on the United States' success in achieving its strategic objectives through a maritime blockade of Cuba during the 1962 missile crisis. This article suggested that 'a maritime blockade in a civil war ... does not come within the scope of international law'. While this argument may find some acceptance in the court of international opinion, it would not be accepted in the United States, where the 1979 Taiwan Relations Act specifically includes a boycott or an embargo among the actions which the United States would consider as a cause for grave concern (Cabestan 2003: 163).

In the eyes of PLA Navy strategists, the legitimacy and appropriateness of naval action to interfere with Taiwan's seaborne commerce would no doubt be enhanced by the memory of Taiwan's actions, supported by the United States, to blockade and harass the Chinese coast in the 1950s. Within a month of Chiang Kai-Shek's (Jiang Jieshi's) retreat from the mainland to Taiwan in May 1949, the Kuomintang government had announced a blockade of all coastal ports in the zone occupied by Communist forces (He 2003: 74). In the immediate aftermath of the Korean War, in October 1953 and again in May 1954, Nationalist warships hijacked China-bound Polish merchant vessels in international waters off Taiwan. In June 1954, a Taiwanese warship seized a Soviet oil tanker making for a Chinese port. In November 1953, a United States National Security Council statement of policy objectives towards Taiwan had advocated 'the increased effectiveness of the Chinese National armed forces ... for raids against the Communist mainland and seaborne commerce with Communist China, and for such offensive operations as may be in the US interest' (Lewis and Xue 1988: 24).

Taiwan's strong dependence on trade and maritime supply makes it inevitable that the Taiwanese economy would be severely damaged by a blockade, or even the threat of a blockade. Shambaugh (2002: 320–1) has observed that a blockade of Taiwan:

> could throw the Taipei stock market and business confidence into free fall and raise insurance rates for freight to unacceptably high levels. During the 1996 crisis, the stock market plummeted 1,000 points (21 percent) and US$15 billion in investment fled the island. The effect on Taiwan's energy needs would be equally devastating, insofar as it is completely dependent on crude oil imports. Taiwan consumes 250,000 barrels of crude every day, and a supertanker docks in Kaohsiung harbour every third day. Moreover, Taiwan is very short on oil: the 120-day strategic reserve built up after the 1995–96 crises had shrunk to a mere 18-day reserve by 1999.

The attraction of blockading Taiwan is all the greater for the PLA Navy because of the limited capability of the Taiwanese armed forces to conduct anti-submarine and anti-mine warfare. Taiwan possesses only two *Guppy*-class submarines of World War II vintage and two ex-Royal Dutch Navy *Zvaardvis* submarines. Taiwan's airborne ASW capabilities are also inadequate. Goldstein and Murray (2004: 180) note a report (dated August 2002) indicating that only six of Taiwan's 26 S-2T Tracker aircraft are operational, a fact that they regard as unsurprising given that these aircraft have been in service in Taiwan since 1976, and were regarded as obsolete by the United

States before that. In their computer-modelled analysis of a hypothetical war between the PRC and the ROC, Rand analysts Shlapak *et al.* (1999: 46) conclude that '*ASW is a critical Taiwanese weakness.* Absent an unexpected acquisition of numerous modern attack submarines, the ROCN will have tremendous difficulty coping with China's modernising submarine fleet.'

Thus, as the Pentagon claims (USDoD 2004b), the balance of naval force across the Taiwan Strait does indeed seem to be moving steadily in Beijing's favour, particularly as Taiwan spends relatively little on its defence and has had only moderate success in reforming its military capabilities, organisation and procedures. Although today the Republic of China Navy (ROCN) still maintains a qualitative edge over the PLA Navy, the gap is closing – among other reasons because, according to the Pentagon's 2002 annual report on Chinese military power, China's large number of submarines could pose a considerable torpedo and mine threat (USDoD 2002). Even if the forces of Taiwan and the mainland would be evenly matched, were they to engage now in a contest for control of the surface of the seas surrounding Taiwan and of the airspace above, China's superiority in the undersea environment could be sufficient to impose or credibly threaten a blockade which would achieve Beijing's strategic objectives vis-à-vis Taiwan.

This task would be facilitated for the PLA Navy if it were also able to secure surface control and air superiority, but it could still be accomplished sufficiently to achieve Beijing's political objectives even in the absence of such control. It should be remembered that Germany did not need command of the sea surface and airspace over the Atlantic for her devastating submarine campaign against British sea lines of communication which came close to bringing Britain to her knees during the Second World War, although increased allied air and surface protection of convoys eventually made German submarine operations more and more difficult (in large part because, until the Type-XXI U-boat became available towards the very end of the war, the German Navy did not have a submarine which could launch attacks effectively while submerged). Nor did the United States enjoy command of the surface to wage a successful submarine campaign against Japanese lines of communication in the western Pacific in the early part of the Pacific War. On the other hand, World War II also demonstrated that effective anti-submarine warfare operations are very difficult, if not impossible, without command of the surface of the sea and the airspace above (Doenitz 1959: 131). In two wars the submarine very nearly proved to be a decisive strategic weapon against Britain. It also came close to inflicting a strategic defeat on Japan in the Second World War. There is no reason to suppose that a Chinese submarine blockade of Taiwan would be any less effective without a significant increase in the anti-submarine warfare capabilities currently available to counter it.

Thus, if Beijing decided to force Taipei by means of a maritime blockade to accept its terms for a political settlement of the Taiwan issue, it could be reasonably confident that the PLA Navy would have the capability, given enough time, to carry out this mission successfully – even without the advanced undersea warfare capabilities provided by its new *Kilo-* and *Song-* class attack submarines.[2] Quoting a 1991 interview with Vice Admiral Zhang Lianzhong, Commander of the Chinese Navy, Lewis and Xue (1994: 228) report that partly against the eventuality of the need to blockade Taiwan, 'the Central Military Commission has ordered the navy to continue constructing submarines and has given "the development of submarines ... precedence over all other [construction]". If another Taiwan Strait crisis should occur, these submarines would represent the front-line force.'

Taiwan's current ASW capabilities would mean that its forces acting alone may not be effective in countering a maritime blockade of the island. The main thrust of the Bush Administration's 2001 arms package to Taiwan, with its eight diesel-electric submarines and 12 P-3C Orion aircraft, was aimed at reinforcing Taiwan's anti-submarine and anti-mine warfare capabilities. However, both these elements of the arms package have run into trouble, with Taipei refusing to accept the American P-3C offer because of its US$4.1 billion price tag, and concern about the cost of buying eight new submarines – even if Taipei and Washington were able to find a country willing to jeopardise its relations with Beijing in order to supply conventional submarines or submarine technology to Taiwan (Agence France Press 2003). Taipei has been reported to have even looked at buying second-hand submarines, including three surplus Russian *Kilos* (Wu 2003). In June 2004, President Putin was reported to have agreed to Russia selling eight *Kilo* submarines to the United States so that Washington could then refit them with American engines and combat systems before selling them to Taipei, although the Taiwan Ministry of National Defence immediately denied this report. The latest twist to this saga at the time of writing (October 2004) was a reported statement by Taiwan's representative in the United States that the eight submarines would probably be built at a shipyard in Mississippi – most likely Ingalls in Pascagoula – as part of the Bush Administration's US$18 billion arms package to Taiwan (Behn 2004). Given that American expertise in the construction of conventionally powered submarines would have to be built up from scratch, and that Taiwanese crews would have to be formed and trained, it is likely that it will take at least a decade before the ROC Navy can deploy these boats. So, while improvements to its airborne and subsurface ASW capabilities still remain uncertain, in the near term Taipei can only hope that some of these deficiencies will be remedied with the entry into service of Taipei's new *Perry-*, *Knox-* and *Lafayette*-class frigates.

If, therefore, Beijing were to try to blockade Taiwan before the ROC Navy's anti-submarine and anti-mine warfare capabilities are sufficiently upgraded to enable Taipei to counter the threat effectively, the United States would come under great pressure to come to Taiwan's assistance. This means that it is not simply the balance of forces across the Taiwan Strait that

Beijing's strategic planners and decision-makers must take into account in calculating the probabilities of success in using military force against Taiwan. Chinese policy-makers know that the balance of force that really matters is not the one between China and Taiwan, but the balance between the PRC's forces on the one hand and those of the United States and its allies on the other. As long as the United States maintains its policy of opposition to the use of force to resolve the Taiwan issue, Chinese strategic planners know that the use of force by Beijing to achieve its political objectives vis-à-vis Taiwan will almost inevitably require Chinese armed forces to challenge the United States successfully for command of the sea and air surrounding the island of Taiwan. Shambaugh (2002: 309) reports an interview with a PLA senior colonel in the Institute of Strategic Studies in Beijing in May 2000, during which he warned that:

> If force is used to reunify [with Taiwan], we must be prepared for war with the United States. If the United States wants to intervene, you are welcome! In order to fight against the United States, the PLA must: (1) maintain a nuclear retaliatory capability; (2) be able to destroy nuclear aircraft carriers; and (3) try to deprive the United States [of the] right to use foreign bases – we shall tell Japan that if they allow the United States to use bases there [in the conflict], we shall strike them!

The deployment of two carrier battle groups to the vicinity of Taiwan during the 1995–6 Taiwan Strait crisis came as a shock to China (Ding 2003: 387). As a consequence, Chinese military planners began to develop various scenarios and contingency plans for US military intervention in future Taiwan Strait crises (Ding 2003: 388). A PLA Navy magazine published an article in 2002 which reveals some of the scenarios for US intervention considered by Chinese strategists. The possible modes of US intervention envisaged by the Chinese range from

> the monitoring of PLA forces, the dispatching of US forces to Taiwan to deter China from escalating the crisis, adopting limited military action to prevent China's military action against Taiwan, and undertaking confrontation actions against China's invading units and logistic units

although this last action is considered unlikely because it could lead to all-out war (Ding 2003: 388). Presumably scenarios and contingency plans have also been worked out for US intervention in the event that China imposes a naval blockade on Taiwan. Chinese military planners would probably consider that the role of US forces in such an event could range from the provision of intelligence, C^3 and technical assistance and equipment to help

Taiwanese forces' anti-blockade actions, mine countermeasures and the formation and protection of convoys to keep open Taiwan's lines of communication, up to punitive strikes against Chinese submarines and their bases, surface ships and aircraft and critical targets such as C^4I units and facilities which would play essential roles in the imposition of a blockade.

The textbook response of a superior naval power to commerce raiders is to neutralise the threat to maritime commerce at source, by attacking enemy ports and bases (Gray 1992b: 14). However, the likelihood of the United States using force directly against the Chinese mainland must be rated as fairly remote. Washington balked at such action even before China acquired its nuclear weapons. Michael McDevitt, formerly Director for Strategy, War Plans, and Policy for the US Pacific Command, has observed that the 50-year reluctance of the United States to use armed force directly against Chinese territory amounts to a strategic tradition: it was General MacArthur's wish to extend the Korean theatre of operations into Chinese territory that provided reason enough for President Truman to dismiss him. This leads McDevitt (2001: 105) to remark:

> I have no particular insight into any US contingency planning on this issue, but were I still the director for strategy, war plans, and policy for the Pacific Command, I would certainly consider in planning for any military intervention in support of Taiwan that land-attack options were off the table, that the only engagements that would be permitted by the National Command Authority would be on, over or under the water.

It is therefore most likely that – at least initially – a conflict directly pitting US armed forces against those of the PLA would be confined to an air and naval battle over, on and under the waters surrounding Taiwan. Even so, as any prudent strategic planners must, Chinese strategists will also have to consider worst-case scenarios, which would include direct action by US forces against mainland targets. PLA strategic planners will therefore have to assume that in a military conflict over Taiwan, US forces will operate in accordance with the doctrine demonstrated in the Gulf War in 1991 and in NATO's Operation Allied Force against Serbia in 1999 by launching intensive precision-guided missile attacks against Chinese air defences, air and naval bases and missile launch sites. Some PLA analysts believe that economic targets, such as oil refineries, transportation hubs and fuel reserves, will also be attacked (Godwin 2003b: 39). There is evidence of PLA preparations for the possibility of air and cruise missile strikes against PLA Navy bases in an exercise conducted by the PLA Navy's submarine force, in which torpedoes were loaded on to a submarine at a small civilian port employing mobile cranes and other special equipment (Goldstein and Murray 2004: 176).

Whereas Beijing could entertain some optimism about the probability of success in bending Taipei to its will by using the PLA Navy to threaten or to actually impose a maritime blockade of the island, as soon as the United States Navy enters into the equation there are no longer any grounds for such optimism. The balance of military forces between China and the United States – not to mention those that could be thrown into the equation by Washington's regional allies, Japan and South Korea – is clearly very much to Beijing's disadvantage and likely to remain so for at least the next two decades. Compared with the United States and its western Pacific maritime alliance partners, China is clearly the inferior naval power. The forward presence of US forces in the western Pacific, together with the American global space surveillance and communications capabilities, ensures that the United States enjoys virtually absolute command of the Pacific Ocean and the marginal East Asian seas and airspace above them. This means that if China is to achieve its political objectives with respect to Taiwan through the use of force – be it an amphibious assault or a maritime and air blockade – it has no choice but to dispute the United States' command of the seas surrounding Taiwan.

But the imbalance of naval forces between China and the United States now and in the foreseeable future is such that prudent PLA Navy commanders could entertain no realistic hope of attempting to dispute US command of the sea by means of a general fleet action. Instead, they are likely to adopt the time-honoured strategy of weak naval powers of seeking to deny command of the sea to the naval forces of the superior power by conducting a war of attrition against US naval forces. The relative strengths of China's armed forces and those of the United States place China in the position described by Corbett:

> where a Power is so inferior in naval force that it could scarcely count even on disputing command by fleet operations, there remained a hope of reducing the relative inferiority by putting part of the enemy's force out of action.
>
> (Corbett 1911: 227)

Since the German submarine *U-21* sank the British cruiser HMS *Pathfinder* in the North Sea on 3 September 1914, providing the first practical evidence of its formidable qualities as means for inferior naval powers to dispute and deny the command of the sea exercised by superior powers (Gilbert 1994: 67), the submarine has been the instrument of choice for this strategic role.

4

SEA CONTROL IN THE
WESTERN PACIFIC

The dominant sea power: United States western Pacific forces – strengths

The United States Pacific Fleet enjoys virtually absolute command of the surface of the western Pacific Ocean and the marginal East Asian seas, and it has the capability rapidly to establish surface control over littoral waters in the region should it wish to do so. The Seventh Fleet, the largest forward-deployed fleet in the United States Navy, is equipped with the some of the most advanced weapons systems in the American naval arsenal. It consists of between 50 and 60 ships, 350 aircraft and 60,000 navy and Marine Corps personnel. The fleet is usually made up of one or two aircraft carriers, three or four cruisers, 18 to 20 destroyers and frigates, five or six submarines, an amphibious command and control ship, five to eight transport and landing ships, 18 logistics and support ships and 16 ships of the Maritime Pre-positioned Force. Its naval air force is made up of 200 aircraft aboard carriers and other ships, 22 land-based maritime patrol aircraft, 10 shore-based utility aircraft and 150–160 Marine Corps aircraft (Weeks and Meconis 1999: 138).

As a legacy of the Second World War and the Cold War, the United States maintains a network of military bases across the Pacific, stretching from the west coast of the continental United States and Alaska in the eastern and northern Pacific, through Hawaii in the central Pacific, to Guam, Okinawa, Japan and South Korea in the western Pacific. Altogether, the United States maintains around 100,000 forward-deployed forces in the Asia-Pacific region.

The US command of the East Asian seas is assured in the first instance by its forward-based forces in the west Pacific region. It is these forces which would be most immediately involved in a conflict with China over the status of Taiwan. The most important US military bases in the western Pacific are in Japan. The home of the US Navy's Commander, Seventh Fleet is at Yokosuka, 48 kilometres south-west of Tokyo. Yokosuka is the most important US naval facility in the western Pacific: it hosts 13 afloat commands and more than 50 other shore commands and activities, including the major naval ship-repair facility in the Far East. The Seventh Fleet's command ship,

USS *Blue Ridge*, together with the US Carrier Battle Group centred around the aircraft carrier *Kitty Hawk*, are home-ported at Yokosuka. Some 9,800 US Navy personnel are stationed there (Weeks and Meconis 1999: 91). Nearby, at Atsugi, is the largest US Naval Air Facility in the Pacific, the home of the US Carrier Air Wing FIVE, the unit assigned to the US aircraft carrier home-ported at Yokosuka. There are over 2,000 US Navy personnel stationed at Atsugi (Weeks and Meconis 1999: 91). On the southwestern coast of Japan are the Sasebo naval base, home to the Commander, US Navy Amphibious Squadron Eleven and six home-ported ships, and the Iwakuni Marine Air Corps Station, home of the 1st Marine Aircraft Wing. Some 5,000 US Navy personnel are based at Sasebo, and around 3,200 active-duty US Marines at Iwakuni (Weeks and Meconis 1999: 92). Misawa Air Base, 644 kilometres north of Tokyo on the northern tip of Honshu, is the home of the US Maritime Patrol air squadron, operating P-3 patrol aircraft and the US Air Force's 35th Fighter Wing, operating F-16 fighters. The US military bases near Tokyo are approximately 1,500 kilometres from the nearest coast of mainland China, and just under 1,900 kilometres from the northern tip of Taiwan. Iwakuni is less than 1,000 kilometres from the Chinese coast and 1,200 kilometres from the northern tip of Taiwan, while Sasebo is less than 700 kilometres from the Chinese mainland and around 1,000 kilometres from Taiwan. Although US land-based fighters could operate over the Taiwan Strait at these distances from bases on Honshu, sortie rates would suffer significant degradation unless substantial aerial refuelling and additional resources and additional crews were deployed (Khalilzad *et al.* 2001: 68).

Closer to a potential combat theatre centred on Taiwan and its surrounding waters are the US bases on Okinawa, where some 28,000 US military personnel are stationed, two-thirds of the total US force in Japan (Weeks and Meconis 1999: 92). The principal bases are the Marine Corps Base Camp Butler and the Kadena Air Base. At Camp Butler, major commands are the III Marine Expeditionary Force, whose primary mission is amphibious assault, and the Marine Corps Air Station at Futenma. Okinawa is also the home of the US Air Force's 18th Air Wing, stationed at Kadena Air Base. Kadena is one of the US Air Force's most important overseas bases, with approximately 7,500 active-duty US military personnel (Weeks and Meconis 1999: 94). The US military bases on Okinawa are only some 500 kilometres from northern Taiwan and just over 600 kilometres from the mainland coast, well within the average operational radius (500 nautical miles/926 kilometres – without refuelling) of current and next-generation US fighter aircraft such as the F-15, F-16, F-22 and F-35 (Khalilzad *et al.* 2001: 68).

In addition to the US Navy and Marine Corps presence in Japan, the US Air Force maintains an important base at Yokota, 45 kilometres northwest of Tokyo, the headquarters of the 5th Air Force, Commander, US Forces

Japan and the 347th Airlift Wing. Yokota is home to some 3,600 US Air Force personnel. The US Army's most important facilities in Japan are at Camp Zama, 40 kilometres southwest of Tokyo, headquarters for the Commander, US Army Japan, and Torii Station on Okinawa, home of the 1st Battalion, 1st Special Forces Group, Airborne (Green Berets), the only forward-deployed Special Forces Unit in the Far East (Weeks and Meconis 1999: 94).

An important contribution to the operations of US forces in wartime operations in an East Asian theatre would be provided by units located on the island of Guam, a United States territory at the southern end of the Marianas island chain, some three hours by air from Tokyo. The US Navy has had three SSNs based at Guam's Apra Harbour facility since the end of 2002 (Global Security.org 2001). Guam is the site of Andersen Air Force Base, home to around 2,200 air force and 460 active-duty navy personnel. Guam is also the location of US Naval Forces Marianas, which comprises some 7,500 active-duty naval personnel. During the Second World War, the Korean War, the Vietnam War and the 1991 Gulf Conflict, Andersen Air Base served as a heavy bomber base (Weeks and Meconis 1999: 90). With its 3,324-metre runway, it could play a similar role in a Taiwan-related war with China. The distance between Guam and the Taiwan Strait is only half the distance between Diego Garcia in the Indian Ocean and central Iraq. RAND analysis indicates that Guam-based B-52 heavy bombers armed with Harpoon anti-ship missiles could play an important role in defeating Chinese maritime operations in the Taiwan Strait (Khalilzad *et al.* 2001: 71). In February 2004, six B-52s were reported to have been transferred to Andersen Air Force Base from their base in North Dakota in response to a request from the US Pacific Command for 'a rotational bomber force on the island as long as it is needed'. Guam is also reportedly under consideration as a base for an aircraft carrier strike group to be transferred from Hawaii (Minnick 2004).

Until the end of 2004, approximately 36,000 United States military personnel were also based in South Korea. In September 2004, the Pentagon announced its intention to reduce this number by some 12,500 over several years from the end of 2004 (Rhem 2004). While the primary mission of these forces is to deter, or if necessary respond to an attack by North Korea against South Korea or Japan, they would also play an important role in the reception, staging, onward movement and integration of the US augmentation units arriving in the Northeast Asian theatre from the continental United States. US Air Force fighter and strike aircraft based at Osan and Kunsan are within combat range of important Chinese surface-to-surface missile bases on the Shandung Peninsula, as well as the PLA Navy's Northern Fleet's main base at Qingdao. The United States also has access agreements which enable its forces to use facilities in Singapore, Malaysia, Thailand, the Philippines and Indonesia.

These forward bases, and the resources located within them, are essential to the United States' strategic presence in the western Pacific. But distances and the maritime geography of the region mean that the operational spearhead of these forces must be provided by the carrier battle groups. The deployment of the USS *Independence* aircraft carrier battle group, followed by the USS *Nimitz* battle group, to diffuse the crisis in the Taiwan Strait in March 1996 was indicative of the critical operational role that these forces would play in an actual conflict.

As former US Secretary of Defense William Cohen reminded Beijing at the time of the 1996 Taiwan Strait crisis, to the great displeasure of the Chinese, the United States has 'the best damn navy in the world!' (Ross 2000: 112n). The United States Navy is without doubt the most powerful navy in the world – in terms of size, reach and resources, equipment, training, experience and professionalism. Although American forward-deployed naval presence in East Asia shrank by about 40 per cent during the decade following the end of the Cold War (IISS 2000: 2), at first glance it would appear that when pitted against the forces of the world's pre-eminent sea power China's relatively under-resourced, technologically backward, inexperienced navy – still in the early stages of its transition from a coastal defence force to a force with wider regional ambitions – would have little hope of success. But the US Navy, for all its formidable qualities, does have a number of weaknesses and vulnerabilities of which Chinese strategic analysts are aware, and which China's strategic planners hope to exploit to their advantage.

United States western Pacific forces – weaknesses

The first of these weaknesses is, paradoxically, a product of the United States Navy's global pre-eminence and the global responsibilities that it has assumed. This means that while the US Navy is by far the strongest in the world and most capable of projecting power in any part of the globe, its global strength does not necessarily translate immediately into regional pre-eminence.

The possibility that regional powers could take advantage of the engagement of US military forces in another region to launch a challenge to US interests has worried American strategic planners since the end of the Cold War. The Clinton Administration's 1993 Bottom-Up Review concluded: 'therefore it is prudent for the United States to maintain sufficient military power to be able to win two major regional conflicts that occur nearly simultaneously' (Weeks and Meconis 1999: 39). Despite this stated intention, doubts persisted throughout the 1990s that the United States would have sufficient military power to win two near-simultaneous major regional conflicts. The Bush Administration's Quadrennial Review, published in September 2001, is more ambiguous – or perhaps more realistic – in this respect, affirming that 'for planning purposes, US forces will remain capable of swiftly defeating attacks against US allies and friends in any two theaters of operation in overlapping timeframes', but then stating that 'US forces will be capable of decisively defeating an adversary in one of the two theaters in which US forces are conducting major combat operations by imposing America's will and removing any future threat it could pose' (USDoD 2001: 21).

The fact is that post-Cold War regional crises have stretched the capacity of US armed forces to the limit, despite the Bush Administration's major increases to the defence budget. In July 2003, stretched by its existing deployments in Iraq and Afghanistan, the Pentagon announced that it had only three combat-ready brigades available for new missions out of a total of 33 (IISS 2003: 12). The US Navy has been no less stretched than US ground and air forces. Regional Commanders-in-Chief maintain that the United States needs, at a minimum, 15 aircraft carriers to implement regional missions (Davis 1998: n. xiv). During the Kosovo crisis in 1999, for example, the navy was forced to redeploy its only aircraft carrier based permanently in East Asia so that it could have one aircraft carrier in the Adriatic and another in the Persian Gulf. To compensate for the absence of the aircraft carrier from East Asia – the first time that the United States had not had an aircraft carrier in Japan for several decades – the Pentagon deployed 40 land-based fighter aircraft to Japan (IISS 2000: 2). More recently, the Pentagon reportedly plans to shift long-range bombers and other aircraft to Guam and elsewhere in Asia to offset the loss of combat power as thousands of US soldiers and marines forward-deployed to that region are redeployed to Iraq and Afghanistan (IHT 2004).

It would be remarkable if, in planning the best time for Beijing to use armed force to coerce Taipei into accepting its political will, Chinese strategists well versed in classical Chinese military thought failed to heed Sun Zi's advice: 'When military weapons are blunt, the morale of the troops is low, the army very exhausted and the supplies of the state are depleted, the neighbouring warlords will capitalise on such misfortunes and vulnerabilities by launching attacks (against you)' (Chow-Hou 2003: 41).

The overstretch evident in the need for US armed forces in the Pacific region to resort to stop-gap measures to make up for extra-regional deployments points to the second of the potential weaknesses of the US forward presence in East Asia. Could those land-based aircraft deployed to Japan during the Kosovo crisis have participated in a US response to a military crisis in the Taiwan Strait, or would the fact of their having to operate from Japanese territory have imposed constraints on their ability to respond quickly and effectively? The bulk of the United States' forces forward-deployed in the Asia-Pacific region are based in Japan. It is on these forces that the responsibility for any initial response to a contingency in the Taiwan Strait would fall. But, depending on the exact circumstances in which the use, or threat of use, of force against Taiwan occurred, the speed and extent to which US forces based on Japanese territory could intervene directly in a conflict involving Chinese armed forces might be circumscribed by Japanese concerns about the impact of such action on Tokyo's relations with Beijing, the possibility of Chinese retaliation against Japan, and sensitivities among the Japanese political elite and general public. Tokyo could decide, for example, not to permit strikes on China itself to be launched from its territory, especially in circumstances where it appeared that Chinese aggression had been prompted by unreasonably provocative behaviour on the part of the Taiwanese authorities. Seoul could apply similar political restrictions to the use of US bases in South Korea for attacks against Chinese territory.

The renegotiation in 1996–7 of the US–Japan Defence Co-operation Guidelines brought to the fore the tensions that exist between Washington and Tokyo over a potential military response by US forces based in Japan to the use of force by mainland China against Taiwan. The guidelines, originally negotiated in 1978 to clarify the role to be played by Japanese defence forces in the event of a response by US forces to a direct attack on Japan's homeland, required updating in the aftermath of the Cold War to enable Japan to play a greater role in responses to regional contingencies. In deference to regional and domestic political sensitivities about any measure which could open the way to Japan's eventual remilitarisation or intervention by its military forces in regional affairs, the revised guidelines deliberately failed to clarify the geographic scope within which the guidelines were to apply by refraining from defining the meaning of the phrase 'areas surrounding Japan' (Reid 1998: 37). Beijing was – and remains – especially sensitive to the implications of the revised guidelines, particularly after a well-publicised comment in August 1997 by Japan's former chief cabinet secretary, Seiroku Kajiyama, that under the guidelines, Japan could provide logistical support to the United States if hostilities erupted in the Taiwan Strait (ibid.: 45). The Japanese government also faced a difficult task in persuading legislators and the public to accept the revised guidelines. They were passed into law only in 1999 after North Korea test-fired a ballistic missile over Japan in late 1998.

Once again, this is a weakness that Chinese strategists and policy-makers would be bound to exploit if they heeded the celebrated advice of Sun Zi: 'The most supreme strategy is to attack the plans and strategies of the enemy. The next best strategy is to attack his relationships and alliances with other nations' (Chow-Hou 2003: 59–60).

The differences which surfaced during the renegotiation of the US–Japan defence guidelines were but one specific instance of a more general concern of the US Navy and other branches of the US armed services about the constraints on their operational freedom of action that result from their reliance on fixed bases in third countries. Operations in Kosovo in 1999 from fixed bases in Italy were subject to restrictions by Italy and other coalition partners. More recently, the United States experienced difficulty in persuading neighbouring states to provide its forces with direct combat support for Operation Enduring Freedom in Afghanistan, and Turkey refused to allow the movement of the US Army's 4th Infantry Division through its territory to participate in Operation Iraqi Freedom. These experiences validated the decision expressed in the September 2002 United States National Security Strategy to restructure and relocate US bases to increase the flexibility of US forces to project power across the world. The US Navy's contribution to this strategy was outlined in its October 2002 blueprint for the evolution of US naval power, *Sea Power 21*. One of the three key concepts in this vision is that of 'Sea Basing', a series of measures to restructure, re-equip and relocate naval forces in order to achieve a greater degree of operational independence through the 'expanded use of secure, mobile and networked sea bases' in the form of 'sovereign platforms operating in the maritime domain'. As US Chief of Naval Operations Admiral Vern Clark (2002) describes it: 'Netted and dispersed sea bases will consist of numerous platforms, including nuclear-powered aircraft carriers, multi-mission destroyers, submarines with Special Forces, and maritime pre-positioned ships, providing greatly expanded power to joint operations.'

Sea Power 21's Sea Basing concept may well give the US forces the kind of operational independence and flexibility that were found to be wanting in post-Cold War conflicts. The shift to greater reliance on mobile, networked sea bases could also help overcome some of the potential constraints on a rapid and effective response to a Taiwan contingency that may face US forces based on Japanese territory. But the Sea Basing concept may not necessarily give forward-deployed US forces the added security that is the second primary motive for the shift to greater reliance on mobile, at-sea platforms as the basis for its operations.

Even before 11 September 2001, the Pentagon worried that the proliferation of weapons of mass destruction and ballistic missiles meant that their forward-deployed forces were increasingly exposed to the risk of an attack by states or non-state actors using unconventional weapons. It had long been assumed that US bases in Japan, Korea and Guam were potential targets for Chinese intermediate-range, nuclear-armed ballistic missiles. From the very start of China's ballistic missile development programme, its missiles were designed with the aim of being able to target US bases in Korea, Japan and the Philippines. The DF-4 intermediate-range missile, for example, was specifically designed with a range of up to 4,000 kilometres to strike the B-52 base on the island of Guam (Gill and Mulvenon 2000: 29). It was also assumed, however, that this risk was covered to a tolerable extent by the United States' own nuclear deterrent force. But the rise of potentially undeterrable actors armed with weapons of mass destruction in the post-Cold War world has greatly increased the Pentagon's estimation of the risk faced by its personnel and assets stationed in fixed overseas bases. The shift to offshore, mobile basing outlined in *Sea Power 21* is a response to this sense of increased vulnerability.

The shift to Sea Basing is thus intended to reduce the vulnerability of US forces to asymmetric attacks by relatively unsophisticated actors such as Kim Jong Il's North Korea or international terrorist organisations. Washington aims to reduce the risk faced by its forward-deployed forces to the kind of unconventional attacks that US Marines suffered in Beirut in 1982 or that US Air Force personnel were subject to in Saudi Arabia in June 1996. But while the use of mobile afloat bases may reduce the risk of terrorist attacks, it could also increase their vulnerability to attack during a more classic type of interstate conflict with the organised forces of a well-armed sea power such as China, a type of opponent that the US Navy has not had to face since the disintegration of Soviet naval power. Bases are vulnerable because they are fixed and therefore their locations are known, and because their location on land makes them relatively accessible to would-be attackers. Mobile, afloat bases overcome these problems. But locating and attacking mobile, at-sea bases would not present an insurmountable problem for a country with the intelligence, surveillance and reconnaissance capabilities, such as China, even if those capabilities were less sophisticated and extensive than those of the United States. And while the mobility of the bases envisaged in *Sea Power 21* would make them less vulnerable to attack by ballistic missiles – unless these missiles were equipped with sophisticated terminal guidance mechanisms – even if they remained beyond the range of China's air and surface forces, they could still be vulnerable to attack by the PLA Navy's submarine force. China's *Kilo*-class submarines have an autonomous range of over 9,000 kilometres which would include all current US bases and facilities in the western Pacific in a radius stretching as far as Guam and Singapore.

Mobile sea bases may still be less vulnerable to submarine attack than US aircraft carriers which venture into the littoral waters of hostile states. The vulnerability of aircraft carriers, which constitute large, flat, difficult-to-manoeuvre and lightly armoured targets filled with combustibles and explosives, has been the subject of debate ever since this category of ship first made its appearance in September 1918 with the completion of the first true aircraft carrier, the British ship *Argus*. In 1944, Bernard Brodie (61) noted that after less than three years of war the British had lost five of the six carriers they had possessed in September 1939, three of them to U-boats; that in six months of war the Japanese had lost the majority of the first-line carriers with which they had begun the conflict; and that the United States had lost four of the seven with which she had started the war. Referring to commentators who had called aircraft carriers the 'clay pigeon of the Navy', he observed that 'there has never before in the history of warfare been a type of vessel that was so large for its time, so important tactically and strategically, and yet at the same time so vulnerable'.

More recently, the aircraft carrier has been criticised for its increasing vulnerability to sophisticated precision-guided, long-range missiles (Friedman and Friedman 1996: 191). This vulnerability first became apparent in the 1960s with the introduction of the Soviet long-range, air-launched land-attack and anti-ship missile, the AS-4 Kitchen. This missile had a range of 300 to 450 kilometres and could be launched by long-range bombers from outside the range of US aircraft carriers' fighter protection screen. It had a speed of up to Mach 3.5 and could dive sharply on to its target at supersonic speeds, making it difficult to intercept (ibid.). The threat to US aircraft carriers posed by the AS-4 Kitchen and its even longer range and more precise successors, such as the AS-6 Kingfish, prompted the development of a single integrated weapons control system to protect carrier battle groups from air, surface and submarine threats. The resulting system was named Aegis (after the shield of the mythological Greek god Zeus) in 1969, and became operational in 1983 when the first dedicated Aegis cruiser, the USS *Ticonderoga*, was commissioned (ibid.: 193). Today, the Aegis system remains the cornerstone for the self-defence of US Navy carrier battle groups.

The anti-submarine warfare component of the Aegis system is provided by hull-mounted SQS-53 active sonars used for both search and attack and SQR-19 passive, towed sonar arrays. The system can also integrate data from other sources – such as reconnaissance satellites, unmanned aerial vehicles or LAMPS (light-airborne multipurpose system) helicopters, which fly off the cruisers' decks and use sonobuoys to detect enemy submarines (ibid.: 195). The radars of the Aegis system can detect targets as far away as 450 kilometres, while the sonars can locate undersea targets tens of kilometres away, depending on water conditions. The latest version of the SQS-53 sonar, the SQS-53C/D, is considered to be capable of detecting the platforms for short-range wake-homing torpedoes before they are able to launch these weapons (FAS 1999). In the case of the UGST wake-homing torpedoes carried by *Kilo*-class submarines, this would mean a range of up to 40 kilometres, although Soviet Type-65 wake-homing torpedoes are thought to have ranges between 90 and 180 kilometres (Luttwak and Koehl 1991: 623).

The first Gulf War revealed that the Aegis system, designed as it was during the Cold War for blue-water operations, experiences performance problems in littoral environments. As George and Meredith Friedman (1996: 198–9) have observed:

[In the Persian Gulf], the clutter from coastlines, pollution from oil fields, electronic emissions from the shore nearby, reduced its capabilities dramatically. Fighters, for example, could not be controlled more than seventy-five miles [120 kilometres] from radar sources ... Aegis is a shield. It can protect a limited territory – defined by the strength of its radar and the range of its weapons. It cannot engage an enemy projectile before it reaches the outer limits of the shield. The faster the projectile is travelling, the less time there is to intercept it; at a certain speed, there is no time at all. Under certain environmental circumstances – such as coming in too close to the shore – the shield contracts. Moreover, the closer in to the shore you are, the more inexpensive launchers will be available to fire projectiles in response. Therefore, the absolute limit of the shield, defined by the number of projectiles available, is compounded by the relative limit – the number of projectiles that can be handled during any one time period.

If there were doubts about Aegis' performance in the Persian Gulf against the relatively poorly equipped, trained and commanded Iraqi forces with their negligible sea denial capabilities, there must be even greater doubt about its ability to provide adequate protection for aircraft carriers operating in the littoral waters surrounding Taiwan, facing the well-armed, well-trained and well-organised forces of the PLA and its large, if relatively less advanced, navy. If the PLA Navy has taken a leaf out of the Soviet book on naval tactics, as it has with so many other aspects of Soviet naval strategy, it may well adopt the tactics exercised by the Soviet Navy in the 1970s and 1980s. These anti-carrier exercises featured orchestrated attacks by submarines, land-based aircraft and surface combatants against enemy surface forces designed to overwhelm the carriers' defence systems (Ranft and Till 1983: 151). There is evidence that PLA Navy strategists may indeed have adhered to the Soviet example: Goldstein and Murray (2004: 193) report that:

> Chinese planners, in the Russian tradition, believe that a carrier battle group can be destroyed with multiwave and multivector saturation attacks with cruise missiles. One recent analysis calculates, 'In order to paralyze a carrier, there must be 8 to 10 direct hits [by] cruise missiles ... and nearly half of the escort vessels have to be destroyed. This ... requires the launch of 70 to 100 anti-ship cruise missiles from all launch platforms in a single attack.' The same analysis describes Russia's Cold War-era 'anti-carrier' forces in great detail and concludes, 'This is Russia's asymmetrical and economical answer to the threat of United States aircraft carriers. In the Russian armed forces, no other force could surely fight this threat except submarines.'

Aegis has never stood the test of intense combat with an adversary, such as China, capable of launching numerous simultaneous attacks from multiple directions by anti-ship cruise missiles launched from the ground, aircraft, ships and submarines, ballistic missiles and torpedoes. Technically the system undoubtedly has a theoretical saturation point, but because in the final analysis it is dependent on its human operators, in practice that saturation point may be considerably lower. The 1987 incident in which human error resulted in the USS *Vincennes* shooting down an Iranian airliner points to a potential cause of failure of the Aegis system. The Iraqi Exocet missile which damaged the USS *Stark* in 1987, killing 37 crew members, is another salutary reminder of the fallibility of ship self-defence systems against such weapons. As the Friedmans (1996: 200) point out, a single error of a defensive system such as the Aegis could have catastrophic, strategic consequences:

> There is a fundamental difference between offensive and defensive weapons systems. The failure of an offensive weapon system represents the failure of the attack but not necessarily of the weapons platform that launched it. The failure of a defensive weapon system represents the destruction of the weapons platform itself. A failure of a Tomahawk missile launched from a Ticonderoga means that an enemy target survives to fight another day; the failure of Aegis could mean that a carrier battle group is destroyed.

Even supporters of the US Navy's new aircraft carrier construction programme concede that these vessels are becoming increasingly vulnerable to attack by sophisticated anti-ship missiles in littoral waters and that, in Jacquelyne K. Davis' (1998: 29) words, 'it may be more prudent for Navy and joint planners to consider, among other things, deployments further from the shore ... the incorporation of advanced (and longer-ranged) strike systems, and enhanced low observability features/techniques'. And, as Ian Storey and You Ji (2004: 88) observe, PLA Navy strategists' study of the vulnerabilities of the US Navy's aircraft carriers may be a factor in the apparent waning of China's interest in obtaining an aircraft carrier capability of its own. According to Storey and You, Chinese proponents of a 'revolution in military affairs' believe that because of their high radar and electromagnetic visibility, and their vulnerability to precision-guided missiles, submarines and mines, aircraft carriers are becoming obsolete. They argue that in modern warfare aircraft carriers have become 'floating coffins'. The ascendancy of RMA advocates in the internal debates about PLA Navy strategy, doctrine and capabilities would certainly provide part of the explanation for the apparent cooling of Beijing's interest in acquiring an aircraft carrier capability for China.

5

MARITIME STRATEGIC THEORY AND THE LOGIC OF CHINA'S SUBMARINE FLEET

Maritime strategic theory provides important insights into the logic of Beijing's decision to invest in the development of a strong undersea warfare capability. Strategic theory aims to distinguish the factors common to all wars – the constants – from what is merely accidental. Like the historian, the strategic analyst is interested less in the unique than in what is general in the unique (Carr 1961: 63). As Mahan, who regarded himself more as a historian than a theorist of sea power, observed (1965: 2),

> while many of the conditions of war vary from age to age with the progress of weapons, there are certain teachings in the school of history which remain constant, and being, therefore, of universal application, can be elevated to the rank of general principles.

Maritime strategic theory is concerned to elucidate, as Corbett (1911: 15) put it, 'the principles which govern a war in which the sea is a substantial factor'. Although their opinions differed on some major principles of maritime strategy, both Corbett and Mahan belonged to the historical school of strategic thought which, in Colin Gray's (1999b: 1) words, 'believes that there are elements common to war and strategy in all periods, in all geographies, and with all technologies'. Corbett would have shared Mahan's view that a principle 'has its root in the essential nature of things, and, however various its application as conditions change, remains a standard to which action must conform to attain success' (Mahan 1965: 7). Corbett would also have agreed with Mahan's assertion that 'the existence of such principles is detected by the study of the past, which reveals them in successes and failures, the same from age to age' (Mahan 1965: 7). Thus this chapter looks into the lessons distilled from past experience with the use of submarines in naval warfare in an attempt to examine in this light the problems and potential solutions facing Chinese naval strategists today.

The classical principles of maritime strategy suggest that a land power which aims to consolidate and extend its influence in a predominantly maritime region at the expense of the pre-eminent sea power must attempt to challenge the dominant sea power for command of the sea. The most decisive way of doing this is to destroy the opponent's fleet in battle or, failing this, neutralise it by means of a blockade. Only if the challenger has the resources, the will and the expertise to construct a fleet capable of doing this can it hope to secure command of the sea by seeking out and destroying the adversary's fleet in a decisive battle – as Rome succeeded in doing against Carthage in the First Punic War, but as Napoleonic France, Imperial Germany and Imperial Japan all tried to do without success. Nazi Germany and the Soviet Union faced a somewhat different but related challenge: with their unfulfilled ambitions to construct major surface combatants for high seas fleets capable of rivalling those of the leading sea powers of the day, they had to find a maritime strategy to secure their regional hegemony knowing that command of the sea was probably beyond their reach. Today, Chinese strategists face a similar challenge. Maritime strategic theory's answer to this problem is to say that a weaker naval force, lacking the capability to secure command of the sea for itself, and therefore compelled to avoid decisive action, can at least hope to achieve its strategic objectives by successfully holding the command of the sea in dispute. As Corbett (1911: 211) described the Royal Navy's pursuit of this strategy:

> The idea was to dispute the control by harassing operations, to exercise control at any place or any moment as we saw the chance, and to prevent the enemy exercising control in spite of his superiority by continually occupying his attention. The idea of mere resistance was hardly present at all. Everything was counterattack, whether upon the enemy's force or his maritime communications. On land, of course, such methods of defence are well known, but they belong more to guerrilla warfare than to regular operations.

Given the United States' ability to assert at will its control of the surface of the marginal seas of the western Pacific and the airspace above them, the only way left open to the PLA Navy to conserve any freedom of action is to operate beneath the surface, beyond the view of American maritime surveillance satellites, aircraft and surface vessels. While the US Navy may be able to establish control over the surface of the China seas – at least beyond the range of shore-based defences – the subsurface is still a no-man's domain. This merely reflects the fact that in the modern era, particularly since the introduction of submarines and mines, absolute sea control is very difficult if not impossible to attain, even in the open oceans. For example, as Milan Vego (1999: 117) points out, 'the Coalition's naval forces attained absolute control of the Gulf during the first few weeks of their war against Iraq in 1991; yet their control of the subsurface in the northern Gulf was disputed because of Iraqi mines.'

More generally, it reflects the pattern established since its first appearance in wartime naval operations of the submarine as the weapon of choice for weaker naval powers to challenge the domination of stronger naval powers. One of the first naval strategic theorists to draw this logical conclusion from an analysis of the experience of the First World War was French Admiral Daveluy. In 1919, Daveluy published a book in which he made an observation drawn from the German experience of 1914–18 that is equally applicable to the strategic circumstances of the PLA Navy almost a century later:

> Surface combatants are indispensable for maritime powers whose aim is to challenge their potential adversaries for command of the sea. On the other hand, nations whose weakness condemns them to be on the defensive at sea have every interest in devoting the entirety of their resources to the construction of submarines; they can obtain no benefit from their surface vessels.
>
> (Coutau-Bégarie 1985: 71)

The submarine – weapon of choice for the maritime underdog

In a memorandum dated 26 November 1941 to the Commander-in-Chief of the German Navy, Grand Admiral Raeder, Karl Doenitz, Commander of U-boats, protested about the diversion of scarce dockyard resources in Brittany from the U-boat arm to the repair of the two battleships, *Scharnhorst* and *Gneisnau*. The torpedo and bomb damage suffered by the *Gneisnau* and the sinking of the battleship *Bismarck* on 27 May of that year, had made it apparent by November 1941 that Germany had lost any hope of wresting control of the surface of the sea from the Royal Navy and its American allies, with whom Germany would be formally at war within the month. Doenitz (1959: 166) argued to his superiors in Berlin that 'dockyard personnel should be employed on the building or repair *only of such vessels as are indispensable to the prosecution of the war*' (Doenitz's emphasis). With reasoning that would strike a chord with PLA Navy strategists today, Doenitz supported his plea by arguing:

> We are in conflict with the two strongest maritime powers in the world, who dominate the Atlantic, the decisive theatre of war at sea. The thrusts made by our surface vessels into this theatre were operations of the greatest boldness. But now, principally as a result of the help being given to Britain by the USA, the time for such exploits is over, and the results which might be achieved do not justify the risks involved. Very soon, as the result of enemy countermeasures, our surface vessels will find themselves compelled to abandon their offensive against the enemy lines of communication in favour of a purely defensive role of avoiding battle with superior

enemy forces ... Only the U-boat, therefore, is capable of remaining for any length of time and fighting in sea areas in which the enemy is predominant, since it alone can still carry out its operations without at the same time being compelled to accept the risk of battle against superior enemy forces. An increase in the number of battleships and cruisers in these waters, far from constituting any addition to the dangers to which the U-boats are exposed, is regarded, on the contrary, as a welcome addition to the targets for which they are searching.

Across the Atlantic, three years after Doenitz wrote his memorandum, Bernard Brodie (1944: 68) made a similar judgement: 'the great strategic value of the submarine lies in the fact that it is the only warship which can operate independently for extended periods in seas which are dominated by the enemy'. The continuing validity of these assessments of the strategic value of the submarine is reflected in the views of contemporary Chinese naval strategists from the Navy Research Institute in Beijing, who have reached a similar conclusion: 'in future naval warfare, the multidimensional battlefield will reveal naval targets and the marine battlefield perspective, making it impossible for surface ships without air force cover to operate in high-threat maritime zones' (Shen *et al.* 1998: 266). These Chinese strategists consider that the application of information technology will make the maritime battlefield so transparent that surface ships and aircraft will be deterred from intervening:

> Such deterrence is multidirectional but much less serious to submarines, because submarines are more difficult to track. Submarines can fulfil combat tasks and attack land targets according to information obtained from the command post while keeping their movement concealed, and they can move under water for a long time without being discovered. The prospect for using submarines is good, because of their covertness and power. Even without attacking targets, submarines are menaces existing anywhere at any time.
>
> (ibid.: 277)

Ever since the first recorded use of a submarine as a warship in 1776, when David Bushnell tried unsuccessfully to use his one-man submersible to sink HMS *Eagle* in New York harbour, the tactical submarine has been traditionally the weapon of the weak against the strong. At the end of the nineteenth century, France, as the principal challenger of British naval supremacy, had become the most advanced designer and operator of submarines. At naval arms control conferences during the 1920s and 1930s, France persistently vetoed British and American proposals for an international ban on submarines. Adopting a stance recalling France's later refusal to ratify the Nuclear Proliferation Treaty during the 1970s and 1980s (a position which incidentally closely mirrored that of the PRC), French delegations opposed the proposed ban on submarines on the grounds that it was the weapon of the weak against the strong and to ban it outright would give an unfair advantage to states which already enjoyed a superiority in surface combatants (Van der Vat 1995: 126).

The potential of the submarine as an offensive weapon was first revealed to the Imperial German Navy, somewhat to the surprise of German naval strategists, during manoeuvres conducted in May 1914 (Castex 1997: vol. I, 278). Once the initial encounters in the Heligoland Bight in August 1914 between the German and British fleets had impressed upon the former the strength of the latter, and once German U-boats had proved themselves in actual wartime conditions by successful attacks against enemy surface combatants in September 1914, the German Naval High Command decided in February 1915 to switch the efforts of its U-boat fleet from a defensive to an offensive role by engaging them in a *guerre de commerce* against enemy merchant shipping in the Atlantic. From this moment the submarine became for the German Navy during both World War I and World War II, in Bernard Brodie's (1944: 169) words, 'more than a naval vessel. It has been a whole theory of naval war':

> Opposed by the greatly superior surface strength of the British Navy, the Germans looked to their submarine fleet as the only means by which they might make a determined bid for victory. Their use at sea of their air force and their surface fleet was lame and inept. All of the tactical and strategic science of which the German mind is capable was lavished on the U-boat. And it did not fail them. It even came close to winning the war for them. That in the end it fell short of doing so by itself can scarcely discredit it as a naval weapon. Supported by or in support of a well-balanced fleet and air force, as the British submarine was in the Atlantic and Mediterranean and as our own has been in the Pacific, the submarine is capable of pulling a good deal more than its own weight in achieving ultimate victory.

The value of the submarine as a means to dispute the local command of marginal seas became apparent even to naval powers who enjoyed general command of the sea. During the First World War the Royal Navy used submarines with significant success in the Baltic Sea and the Sea of Marmara where the German and Turkish navies enjoyed local predominance. During the Second World War, after the British had lost command of the central Mediterranean, the Royal Navy used its submarines to deny the use of the sea for Axis communications between Europe and North Africa (Hezlet 1967:

160). American submarines operated with significant success against Japanese shipping in the western Pacific in the months before the US Navy had wrested control of those seas from the Imperial Japanese Navy. Once the United States had recaptured Guam and Saipan in the summer of 1944, submarines based on those islands imposed a virtual blockade on Japan. It is widely contended that this submarine blockade would in itself have brought Japan to capitulate without the need to subject Japanese cities to incendiary and nuclear bombardment (Blair 1975: 17). Hezlet (1967: 209) maintains that:

> American submarines in the Pacific in World War II were one of the most important factors in the defeat of the Japanese Navy. They were of more value than a battlefleet and in the same class as carrier-borne aircraft, even when employed part-time against warships.

Germany and China: disputing sea command by turning weakness into strength

Although China has an avowed aim of developing a blue-water navy by the middle of this century, its maritime strategy for now can only realistically be based on the fact that its sea power is weak relative to that of its principal potential adversaries. For at least the next 10 to 15 years, the reality for China's strategic planners is that its naval power is that of a coastal state confined to the narrow seas enclosed by the first island chain along the edge of the Asian continental shelf. Chinese naval strategists must nonetheless plan for conflict with the world's leading naval power. These factors establish the basic parameters within which Chinese strategists must work for the time being.

Strategy is about ends and means. The art of strategy, according to General André Beaufre (1963: 36), is to select from among a range of possible instruments – from nuclear bombardment to propaganda to trade agreements – and combine them to produce the desired effect on the adversary. The choice of means to achieve this end will be determined by the confrontation between one's own possibilities and the vulnerabilities of the adversary. The choice of strategy as a function of the relationship between one's own means and those of the adversary is one of the essential tasks of the strategist. Castex (1997: vol. II, 1) quotes an analogous remark by his earlier contemporary, Daveluy: 'strategy is characterised by making the most of limited resources, or at least by ensuring that one's resources provide maximum value'. The key to the choice of strategy is the assessment of the balance of strength in the adversarial relationship.

For continental Germany in 1917 and again in 1942 and in early 1943, as Gray (1990: 63) notes, the intended instrument for decision in war was the U-boat – a choice indicating that 'the relative importance of a kind of military power is a function of a particular vulnerability of the enemy of the day'. Similarly, Beijing's choice of submarine warfare as a primary instrument of its military strategy is the product of its assessment of its own forces' strengths and weaknesses and those of its potential adversaries – Taiwan and the United States and its allies. These equations of force interact with other objective factors including geography, technological competence and financial resources, as well as the more subjective factors which comprise China's strategic culture, to make the development of a strong undersea warfare capability a logical choice for China's strategists. An effective submarine force plays to the strengths of China's position and strategic traditions and, at the same time, exploits perceived weaknesses in the forces of China's potential adversaries. Chinese naval strategists are well aware of the relevance of the German experience in submarine warfare to the PLA Navy's contemporary tasks. Goldstein and Murray (2004: 175) have noted an increase in detailed discussion of the Battle of the Atlantic during World War II in Chinese military periodicals, and that undersea warfare figures prominently in these analyses. They observe that 'Adolf Hitler's U-boat campaign is of great interest to Chinese strategists, as is Germany's broader evolution as a maritime power'.

Admiral Raoul Castex was the first (and some would argue the last)[1] important theorist to analyse systematically the implications of submarine operations during the First World War for maritime strategic theory. In 1920 he published his *Synthèse de la guerre sous-marine*, one of the first studies to focus on the implications of this new weapon for naval strategy (Coutau-Bégarie 1985: 86). Castex's ideas on submarine warfare were translated into German and are believed to have greatly influenced German planning for the submarine campaign in the Second World War (Brodie 1944: 296). Castex's ideas about the use of the submarine as a commerce raider were reflected, for example, in Karl Doenitz's book, *Die U-bootswaffe*, published in 1939 (ibid.: 87).

The paramount lesson of the First World War for Castex was that the submarine, by itself, could not be an instrument with which the inferior naval power could hope to wrest the command of the sea from the superior power. In other words, the submarine was a weapon for sea denial, not sea control. It could not therefore be a decisive weapon:

> In the final analysis, the result was that the submarine, acting alone, without the support of surface forces, even employed as it was under conditions which were so favourable to its use that they are never again likely to be repeated, revealed itself to be powerless to win command of the sea ... It was unable to ensure for itself the control of essential surface communications. Neither was it able to seize this control from the enemy.
>
> (Castex 1997: vol. I, 293)

With the benefit of the experience of the use of submarines during World War II, Karl Doenitz would have disputed Castex's verdict on the submarine's incapacity to be a decisive weapon, even though he may have agreed that the submarine, by itself, cannot secure command of the sea. Doenitz believed that 'after three and a half years of war [Germany] had brought British maritime power to the brink of defeat in the Battle of the Atlantic – and that with only half the number of U-boats which [he] had always demanded'. In Doenitz's view, Germany lost the war because the continentalist-minded German leadership had believed that the war, in which Germany's main opponents were the two greatest sea powers in the world, could be won on land and had therefore starved the U-boat arm of the resources it needed to prosecute the war at sea successfully (Doenitz 1959: 333). Sir Arthur Hezlet (1967) also argues convincingly that the use of submarines in World War II clearly demonstrated their potential to be a decisive weapon of sea power. According to Hezlet (ibid.: 106–7), Doenitz's submarines demonstrated that the *guerre de course*, contrary to all the teaching of the past, could be 'a potentially decisive method of waging war. There is little doubt that, unlike all wars of attrition in the past, [the submarine] could now not only dispute but also prevent another power from exercising command of the sea.'

Castex nevertheless recognised that the submarine, if exploited to its maximum advantage, could be a very powerful weapon in the hands of a power too weak to challenge the dominant naval power's command of the surface. The submarine could give the side, whose inferiority on the surface would have condemned it in earlier days to a passive form of defence, the means to conduct offensive operations: 'the submarine, like the aeroplane, is an active instrument of defensive offence, suited to the practice, in cases of too great an inferiority on the surface, of stubborn attrition warfare, of these "minor counter-attacks" of which Corbett speaks' (Castex 1997: vol. I, 318).

In the light of certain structural similarities between the situation faced by the Imperial German Navy in the First World War and that facing the PLA Navy in a potential conflict with the United States Navy during the coming ten or fifteen years, Castex's analysis of the German High Seas Fleet's implementation of the operational concepts worked out by its Commander-in-Chief, Admiral Scheer, is particularly interesting.

Like the High Seas Fleet in 1914, bottled up in the narrow North Sea by the superior British Grand Fleet guarding the entrances to the open Atlantic ocean from its bases in the English Channel, Scotland and the Orkney Islands, the PLA Navy's freedom of action is circumscribed by the superior naval power's control of the exits from the narrow China seas into the open Ocean. In the event of a conflict over Taiwan, the only way for Chinese warships to pass from their bases into the wider Pacific through straits not controlled on both sides by potentially hostile forces is through the southern end of the Taiwan Strait, and even then they would still have to pass through the deep but relatively narrow Luzon Strait between Taiwan and the Philippine islands. In such a conflict – as discussed above in Chapter 2 – Chinese naval strategists would no doubt expect US and Japanese forces to adopt tactics similar to those used during the Cold War against Soviet submarines by creating a barrier of fixed ASW sensors and weapons, tactical submarines, surface anti-submarine ships and ASW aircraft across the strategic straits leading to the open ocean. In the case of China, such a barrier could be interposed across the strategic straits linking Taiwan with the Ryukyu island chain and the Japanese home islands. However, the effectiveness of such barriers to prevent the passage of advanced, stealthy diesel-electric submarines such as the *Kilo*-class is an open question. As Castex (1997: vol. I, 297) observed, the submarine offers an 'extraordinary aptitude' for attacking lines of communications in situations where naval forces have to act in the rear of enemy forces despite the obstacles to such action erected by the enemy.

Castex's analysis of the German predicament during the First World War was validated by Doenitz's experience as Commander of U-boats during the Second World War:

> For the U-boat the strategic disadvantages of our geographical position in relation to England is of very much less importance. First and foremost, the U-boat can submerge. It can therefore pass unseen and, in principle at least, free of the necessity of having to fight enemy surface ships contesting its passage, through the North Sea and out on to the British lines of communication across the Atlantic. For the same reason, and thanks also to the exceptionally large radius of action and its ability to keep the sea far longer than surface vessels, it can remain for longer periods in the strategically important areas. For these reasons and apart altogether from those characteristics which make it an ideal weapon of offence, it was the most suitable weapon for direct attack on Britain's supply lines and for the achievement of our naval strategic aims. In this way the U-boat came very near to costing Britain the First World War.
>
> (Doenitz 1959: 40)

During the 1914–18 war, German submarines had been used predominantly as a means of attacking communications in an indirect strategy, and relatively little as a warship in a direct strategy of strikes against the organised forces of the adversary. Castex believed, however, that the Germans, in concentrating their submarine warfare campaign against Allied merchant shipping, had neglected important lessons from the period between the outbreak of the war in August 1914 and early February 1915, and a subsequent period in the first half of 1916 up until the Battle of Jutland, when Germany employed her submarines in a direct strategy of strikes against the enemy fleet in an attempt designed to equalise the balance of strength between the opposing fleets. He considered that the German Naval High Command had not drawn the right lessons from their submarine fleet's promising debut as a strike force against enemy warships. Had they been more persistent in pursuing the concepts for joint operations between surface and submarine forces developed by Admiral Scheer when he took command of the Imperial German Navy in 1916, they would have made better use of this 'newcomer' to naval strategy. At a time when Chinese naval strategists are contemplating the odds of inflicting significant damage on the capital ships of the United States Navy in the narrow seas off the China coast, Castex's analysis of Scheer's operations against the Grand Fleet in the shallow, confined North Sea in 1916 has some contemporary relevance.

According to Castex (1997: vol. I, 301), Scheer had concluded that despite some notable successes, such as the attack which resulted in the destruction of three cruisers (*Aboukir*, *Hogue* and *Cressy*) on 22 September 1914, and the later sinking of the *Hermes* and the *Attentive*, submarines acting as lone raiders had not been particularly effective in inflicting damage on enemy surface combatants. Scheer realised that better results could be obtained by manoeuvring the German surface fleet in such a way as to draw the large enemy surface ships towards the submarines. Scheer's concept was put to the test in operations in April and at the end of May 1916 – in the lead-up to the Battle of Jutland – in which surface combatants and submarines worked closely with one another. In the first case, the operation yielded meagre results because only two of the 11 submarines which were to have participated in the operation were able to do so. In the second case, where the German submarine flotillas had been positioned outside the British bases to ambush the Grand Fleet when it came out, communications failures due to the primitive radio technology then available undermined the effectiveness of the operation.

Reflecting on the lessons of this experience, Scheer then devised an alternative operational concept in which mobile barriers of submarines would be placed across the enemy's probable route, not close his bases, but near the bulk of the German High Seas Fleet (ibid.: vol. I, 303). The idea was to provide a tempting opportunity for a fleet action to draw the British capital ships into an ambush. This concept was applied during an operation on 19 August 1916. Two lines of U-boats were positioned to lie in wait in the approaches to the British bases at Sunderland and Flamborough. They succeeded in sinking two cruisers, but Scheer realised that better results could have been obtained had the U-boats been positioned in a triple row, with each boat in the rear lines covering a gap in the line ahead. Scheer was not able to refine the tactics for this direct strategy any further, however, since on 7 October 1916 it was decided to divert the High Seas Fleet's submarine arm from the North Sea to the Atlantic to resume the *guerre de course* against Allied merchant shipping (ibid.: vol. I, 302).

Castex considered that the German experience in the North Sea in 1916, inconclusive though it had been, revealed the potential effectiveness of joint surface–submarine operations as part of a strategy to engage the adversary's fleet directly. He thought that it would be possible, in the way Scheer planned but was never able to put into practice, to form groups of submarines which would maintain a certain distance from the surface combatants but still be able to act in concert with them, by positioning themselves along enemy lines of attack or retreat.

Although the tactical results of the action in August 1916 were no doubt disappointing to Admiral Scheer, he probably achieved far more than he realised at the time. The conclusions reached about the strategic impact of this action by Hezlet (1967: 72) reveal much about the strategic value of submarines as a means for inferior navies to deny command of the sea to the dominant sea power:

> The psychological effect of the U-boats was, however, far greater than the damage they did or their value for reconnaissance. Admiral Jellicoe realised that the whole point of the German operation had been to lure him into a submarine trap. He believed that the Grand Fleet had been in great peril and he was determined that it should not be risked in this way again. In other words, the Grand Fleet must avoid action with the High Seas Fleet when their aim was to ambush him with U-boats. Mines had already made it too dangerous for the Grand Fleet to enter the area south of the latitude of Horns Reef and east of 5°E. Admiral Jellicoe now reckoned that he ought not to go south of 55°30' N or east of 4°E at any time, and only west of the same longitude if a really good chance of action with the High Seas Fleet presented itself. Furthermore, he did not believe that he should go south of the Dogger Bank at all unless all classes of ship, including light cruisers, had destroyer screens. As he knew he had not enough destroyers for this purpose, it was tantamount to saying that the Grand Fleet would in future have to stay north of the Dogger Bank.

In other words, Hezlet concludes (ibid.: 73), Scheer's tactical use of submarines had cost the Grand Fleet its command over half the North Sea, including much of the British coastline. Moreover, as Hezlet goes on to point out, the German submarines had the effect of denying this vast area to the Grand Fleet whether they were there or not. In fact, these restrictions on the Grand Fleet's liberty of action remained even after the U-boats had been redeployed to their *guerre de course* against merchant shipping.

During the Second World War, Doenitz, firmly convinced that his U-boats should be used as a strategic weapon against enemy merchant shipping in the open ocean, did not seek to exploit the lessons of Scheer's experience in using submarines as a weapon of attrition against the enemy fleet, although he did adopt the idea of groups of submarines acting in concert, rather than as lone raiders. The true inheritor of Scheer's experience was the Soviet Navy which, in the 1950s and 1960s, developed operational concepts to implement a strategy of sea denial against Western carrier strike groups. The instruments of choice for this sea denial strategy were submarines, raiding warships, land-based aircraft, inshore combatants and mine-warfare vessels (Ranft and Till 1983: 145). As we shall see when we come to examine the influence of the Soviet model of naval strategy and force structure on the PLA Navy, the Chinese may also be among the principal beneficiaries of the lessons learned by the Germans in the North Sea in 1916.

Reflecting on the general lessons to be learned about the submarine as an instrument of naval power from the Germans' experience of U-boat operations in the North Sea during the First World War, Castex (1997: vol. I, 319), steeped as he was in the strategic tradition of the previous two centuries' Anglo-French naval wars, concluded somewhat ruefully:

> The submarine is, like the aeroplane, an active instrument of offensive defence, perfect for the practice, in cases of too great an inferiority on the surface, of that war of attrition, the 'minor counter-attacks' of which Corbett speaks. The Germans, almost to a fault, clearly showed the method and the way to follow in similar cases, and our ancestors of the Revolution and Empire would have given a lot to put their hands on such a tool.

Revolution in naval affairs – the submarine as an asymmetric weapon

His analysis of these German undersea warfare operations led Castex to the conclusion that although the advent of the submarine as a new weapon of naval warfare did not mean that it was about to replace surface fleets, it nonetheless represented a revolution in maritime strategy. It meant that henceforth, under the influence of this new weapon, command of the sea had become merely command of the surface of the sea (ibid.: vol. I, 307) – a conclusion also reached by Brodie. Although the threat posed by the submarine was not altogether new in 1914 – the torpedo boat had posed a similar kind of threat – according to Castex (ibid.: vol. I, 310):

> The risk from submarines and mines, invisible, difficult to avoid, almost impossible to overcome by any immediate offensive reaction, is altogether more formidable. And if one might have believed, before the last war, that one would only have to deal with them near the coasts, we now know that submarines will be found everywhere and that seabeds susceptible to mining extend much further seawards than before. Nonetheless, there will always be regions where the danger will be greater than elsewhere and where the undersea terrain will be particularly treacherous and sown with

obstacles, such as certain areas of the North Sea during the last war ...
In short, today, submarines and mines result in a significant restric-
tion of one's freedom of action, this indispensable requirement of
the strategist, and put him in a much less favourable position than
he enjoyed formerly.

Half a century before the concept of asymmetric weapons and strategies
acquired currency in strategic thought, Castex described the effect that the
submarine had in expanding the space and time in the maritime theatre of
operations. In the hands of a weak navy confronting a dominant sea power,
this effect created a force-multiplier out of all proportion to the resources
invested by the weaker power in its submarine force relative to the resources
required by the stronger power to neutralise the threat posed by those
submarines to its surface assets.

Submarines have this force-multiplier effect because, according to Castex,
in the days before submarines became operationally effective, in any particular theatre, the areas of tactical interest were restricted to those points
where it was possible to meet the enemy's surface units. To achieve its
maximum effectiveness in an engagement with an enemy, a battle fleet must
concentrate. This principle of concentration means that surface warships
tend to mass within a restricted area, precipitating a corresponding concentration on the part of the opposing forces. To do otherwise would risk the
destruction in detail of the dispersed fleet. In the days where fleets were
composed only of surface combatants, this double concentration had the
effect of reducing the areas of possible convergence and engagement to a
few isolated points of tactical influence scattered like islands across the vast
expanse of ocean. In between these islands of tactical influence stretched the
immense 'blank' areas which were properly the province of strategy (ibid.:
vol. I, 12). Brodie (1944: 136) provided a graphic illustration of this contrast
between the tactical – as opposed to the strategic – area of influence of a
surface fleet when he wrote:

> If one touched the point of a blunt pencil to the chart of a large
> maritime region like the North Atlantic or the Eastern Pacific, one
> would have darkened an area large enough to contain in cruising
> formation the major ships of the largest fleet in existence today. It is
> phenomenal enough that a battle fleet occupying at any one time so
> tiny an area should exert as great an influence over vast seas as it
> does.

The tendency of tactical areas of influence to reduce to a limited number
of points was reinforced, according to Castex (1997: vol. I, 13), by the transition of ships' motive power from sail to internal combustion. The need to
refuel greatly restricted the radius within which machine-age navies could

exert their power. Submarines, however, were not subject to the same restrictions as surface ships:

> The situation of the submarine is completely different. It can be dispersed without risk. It has no fear of isolation or of being left to its own devices. Its invisibility and the very great probability of finding a safe retreat in the depths enable it to envisage without too much concern a counteroffensive which is the sentence of death for surface units left alone to fend for themselves ... Moreover, the submarine is less subject than surface ships to the constraints of resupply, its very large range enabling it to undertake long cruises and greatly enlarging its area of operation ... [This means that] the submarine has no reason to mass systematically in tightly formed groups occupying only a small area of the theatre of operations.

These qualities make the submarine a very effective force-multiplier. In the eyes of the adversary, the submarine's invisibility gives it the quality of ubiquity. Thus, in the presence of a submarine threat, the adversary has no choice but to adopt permanent defensive measures against the possibility of an encounter with submarines at any location in the potential combat zone. This has the effect of extending the areas of tactical interest to cover the whole theatre of operations. Owen Cote (2003: 83) has noted the similarities between the challenge to a superior naval forces posed by submarines and that posed by Iraqi Scud missiles during Operation Desert Storm:

> Operationally, Scud hunting was like ASW against a quiet target. A large area needed to be searched for objects that easily blended into the background and only intermittently exposed themselves. Thus radar was used to flood Scud operating areas, unattended field sensors were also deployed, and aircraft were used to pounce on potential contacts. This was a protracted, extremely asset-intensive endeavor, characterized by false alarms, high weapon expenditures, and low success rates. In short, a Scud launcher was most likely to reveal itself by successfully launching its weapon, just as sinking ships are often the only reliable indication that there is a submarine in the neighbourhood.

Castex considered that the submarine had the effect of an expansion not only of the space in which combat could potentially occur, but also of the time. In the days when naval engagements involved only surface vessels, actual combat occurred sporadically, interspersed by sometimes long intervals. But the submarine's ubiquity in space also implies a continual presence in time, at least in the mind of its opponent. Its ability to strike suddenly anywhere also means an ability to strike at any time. Anti-submarine defences and countermeasures must therefore be maintained in a constant state of readiness for instant reaction (Castex 1997: vol. I, 15).

By dividing and dispersing the stronger naval power's forces, this asymmetric weapon effectively multiplied the weaker power's forces. Castex's analysis of the way that the submarine had produced this effect in the First World War was echoed by Brodie in the latter's analysis of the strategic effects of submarine warfare during the Second World War. Brodie (1944: 75) considered that 'measures for the defence of shipping against submarines generally absorb military resources totalling far more "units of war power" than the enemy puts into his submarine offensive'. A further achievement of a submarine fleet, according to Brodie, was to make 'the adversary disperse his force and cause a strain upon his naval resources which may endanger his whole strategic position' (ibid.: 75). Gorshkov (1979: 120) noted that 'against the German submarines operated 5,500 specially constructed anti-submarine ships and 20,000 small craft. To every one German submarine at sea were 100 English and Americans.' He quotes American estimates that the cost to the Allies of the capabilities to combat submarines in the First World War was 19 times Germany's expenditure needed to build those submarines (Gorshkov 1979: 259).

By February 1943, it had become apparent to the German Naval High Command that its U-boats could not hope to sink more ships than the Allies could replace with newly constructed tonnage. Even though they realised that they had lost the Battle of the Atlantic, the German Naval Staff nevertheless determined that there should be no let-up in the U-boat war because the opportunity costs that it imposed on Allied resources would continue to be an important contribution to the overall war effort:

> Of even greater consequence are the vast resources in men and material which the enemy has devoted to the air and sea formations engaged in combating the U-boat. If the menace constituted by our attacks on tonnage were removed, a war potential of quite unpredictable strength would be released for action elsewhere … Even if the U-boat arm finds itself unable completely to overcome current difficulties and to maintain the successes of the past, it must nevertheless continue the fight with all available forces, since by their activities the U-boats destroy or tie down enemy forces many times their own strength.
>
> (Doenitz 1959: 344)

Doenitz was not mistaken. Hezlet (1967: 102) calculates that by the end of the war, the Germans had 179 U-boats in commission, 121 of which were operational and 58 engaged in trials or training. The boats were manned by some 13,000 men, while another 113,000 personnel were engaged in related training, overhaul, repair and building work. Against these resources, the

Allied side committed over 300 destroyers, sloops and P-boats exclusively on anti-submarine work, as well as 35 submarines, 550 aircraft, 75 airships and nearly 4,000 auxiliary vessels. This force engaged around 140,000 men directly, and probably another half a million indirectly in construction, refit and general support.

'The ideal tactical weapon of offence'

In a 1937 addendum to his *Théories stratégiques*, Castex considered that in his earlier text he should have given greater emphasis to his opinion that the submarine should be used primarily as an offensive weapon:

> As much as possible, the use of submarines must be avoided in essentially defensive, inert, static, geographically fixed dispositions, such as stationed before enemy bases, passive area surveillance, barrages across straits or narrow inlets, etc. ... [Submarines should be] employed against adversary forces and interests in offensive, mobile, dynamic manoeuvres, maintaining constantly the operational initiative, and above all where the submarines have the greatest chance of meeting the enemy, in other words near the point to which he is drawn (the bulk of our forces, convoy escorted by them, etc.). That is where the worthwhile missions will be found for which there are never enough resources.
>
> (Castex 1997: vol. I, 320–1)

Like Castex, Doenitz (1959: 11) also considered the submarine to be 'ideal as a tactical weapon of offence'. This was particularly so for a state like Germany which was strategically handicapped by geography. Hezlet reaches similar conclusions about the defensive versus the offensive use of submarines. During the First World War, he writes (1967: 80),

> although a considerable proportion of British submarine strength was allocated to the defence throughout the war they proved a complete failure in this role. In all the raids by the German fleet on the east coast of England not a single ship was intercepted.

The relative weakness of the submarine as a defensive weapon was confirmed by the American experience in the Pacific during the Second World War. The 29 submarines of the American Asiatic Fleet based in the Philippines, the more modern and powerful of the two US submarine forces in the Pacific, constituted the main naval defence for the Philippines. When it came to the crunch, however, 'they proved of very little value' (Hezlet: 1967: 194). This experience confirmed for Hezlet (ibid.: 194–5) that: 'Submarines have never proved a good weapon in defence: it is always extremely difficult to put them in the right place and their strategic redeployment to meet enemy moves is invariably too slow.'

Its qualities as an offensive weapon, however, make the submarine the ideal instrument for the naval power which, because of the inferiority of its forces, is obliged to go on to the strategic defensive. For Castex, the worst that a naval power forced on to the defensive could do was to adopt a passive defensive posture. Citing Jomini and Clausewitz in support of his argument, Castex (1997: vol. IV, 131) recommended instead what he called, for want of a better description, the 'defensive-offence':

> The side whose serious inferiority on the surface condemns it to the defensive ought always, despite its unfavourable situation, to try to be as active and aggressive as possible. Its fleet should remember that the very fact of its existence is sufficient to confer upon it the title of 'fleet in being', and that, if it wants to have some influence on events, it has to give some sign of its state of being, which it can do by undertaking something, by trying to impose its will to the extent that its means allow, seeking as far as it can the operational initiative, even if it does not result in anything decisive.

In naval warfare, the aim, according to Castex, is to prevent the enemy from the enjoyment in peace of his control of communications. Like Corbett, Castex considered that while on land a defensive posture evoked principally the idea of organised and fortified positions, in naval warfare the principal defensive concept was that of the activity of the inferior side. It meant refusing to engage in a decisive battle, but instead constantly harassing the enemy by limited actions, anywhere and everywhere a favourable opportunity could be found: 'Naval defence should be an unceasing will to counter-attack, alert for every occasion and exploiting them without delay, by every means while, on the other hand, knowing how to stop when the conditions are no longer favourable' (ibid.: vol. IV, 133). In fact, Mahan had anticipated the views of both Corbett and Castex by his formulation of the more general and fundamental principle of all naval war, namely 'that defence is only insured by offence' – a principle regarded by Herbert Rosinski (1939: 1) as the keystone of Mahan's theory of naval strategy, and the essential difference between war at sea and war on land.

Hezlet's conclusions on the value of the submarine when used in an offensive role during the two World Wars are even more trenchant than those of Castex and Doenitz. As a commerce raider in both wars it had proved to be a potentially decisive weapon. As warships used directly as weapons of attrition against the enemy fleet, their record was impressive. During the First World War, although German U-boats were used predominantly in the *guerre de course* against Allied commerce, they still managed to sink six battleships, 10 cruisers, 42 smaller warships and 66 auxiliaries, including 11 armed merchant ships (Hezlet 1967: 82). The U-boats' failure to sink any *Dreadnought*-class battleships, the type that really mattered, rather than older, slower and less well-armoured warships, suggests that the submarine had not, by that time, evolved sufficiently to become decisive as a weapon of attrition.

It was during the Second World War that the submarine came into its own in this role. In the Atlantic, the submarine's performance as a military weapon – as opposed to a commerce raider – was diminished because neither side deployed many of these vessels and because the Germans used the U-boats mainly to attack commerce. But in the Pacific, where both the Japanese and the American pre-war submarine training had focused on offensive operations against enemy warships, the submarine scored some significant successes. Japanese submarines were particularly successful in naval actions around Guadalcanal during the second half of 1942. Without suffering any loss themselves, against heavily escorted American task forces, Japanese submarines managed to sink the aircraft carrier *Wasp*, damaged the *Saratoga* and narrowly missed the *Hornet*. They damaged the battleship *North Carolina* and the cruiser *Chester*, and sank the light cruiser *Juneau* and the destroyer *Porter* (Hezlet 1967: 198–9). The principal role for American submarines, contrary to pre-war expectations, was as commerce raiders. But in the course of their patrols against Japanese merchant ships in 1944 alone, they also accounted for numerous Japanese warships, including the aircraft carrier *Shimano* and the smaller aircraft carriers *Unryu*, *Otaka*, *Unyo* and *Jinyo* as well as six light cruisers and around 30 destroyers (ibid.: 204). Altogether, American submarines destroyed just under a third of the Japanese Navy (ibid.: 207).

6

GEOGRAPHY, NARROW SEAS AND SUBMARINE TERRAIN

Coastal power versus blue-water fleets

So long as it has no direct access to the open Pacific Ocean, China, like Germany, is handicapped by geography. 'Geography,' as Colin Gray (1999a: 165) argues, 'is inescapable.' Geography not only provides 'the physical playing field for those who design and execute strategy' but also 'drives, certainly shapes, the technological choices that dominate tactics, logistics, institutions, and military cultures'. For over a century, since the Japanese annexation of the Ryukyu Islands in 1879 and Peking's cession of Taiwan to Japan under the Treaty of Shimonoseki in 1895, a fundamental geopolitical reality for Chinese naval strategists is that the loss of Chinese control over the island chain along the edge of the continental shelf enables hostile powers to establish a distant blockade, thereby containing Chinese fleets and Chinese offensive sea power within the confines of the narrow China seas.

So long as the first island chain is in the hands of forces opposed to those of the PRC, the PLA Navy will face a formidable obstacle in securing a working control of the sea beyond that barrier which would permit Beijing to project power outside China's littoral waters. The PLA Navy thus has little choice but to adopt a defensive strategic posture because it does not have, in Corbett's words (1911: 35), 'sufficient strength for offence'. It is therefore geography as much as any economic or technical factor that obliges the PLA Navy to adopt a generally defensive posture at the strategic level (see Chapter 2). However, Beijing receives some measure of compensation at the operational and tactical levels for the weak hand that geography has dealt it at the strategic level. Although the PLA Navy's overall strategic position may be defensive, this does not mean that it is weak; nor does it mean that it is condemned to a defensive posture at the operational and tactical levels. Indeed, to paraphrase Corbett (1911: 35), even though the PLA Navy does not have the strength for the offence, the defence gives the PRC 'a special strength for the attainment of its object'. Corbett's explanation of the advantages of a defensive position points to the essence of the PLA Navy's strategy and to the logic of its choice of tactical submarines as the key instrument for its execution:

> If either by land or sea we can take a defensive position so good that it cannot be turned and must be broken down before our enemy can reach his objective, then the advantage of dexterity and strength passes to us. We choose our ground for the trial of strength. We are hidden on familiar ground; he is exposed on ground that is less familiar. We can lay traps and prepare surprises by counter-attack, when he is most dangerously exposed. Hence the paradoxical doctrine that where defence is sound and well designed the advantage of surprise is against the attack.
>
> (ibid.)

The strategic defensive, explains Corbett (ibid.: 38), extrapolating from the experience of the Royal Navy in centuries of conflict against Britain's continental rivals,

> usually meant that the enemy remained in his own waters and near his own bases, where it was almost impossible for us to attack him with decisive result, and whence he always threatened us with counter-attack at moments of exhaustion.

Writing at a time when orthodox military strategy was dominated by the doctrine of the offensive and the widespread belief that the defensive at sea was a strategy of the weak and pusillanimous (a 'pestilent heresy', as he put it [ibid.: 92]), Corbett (ibid.: 32) was at pains to insist that 'defence is not a passive attitude, for that is the negation of war'. He characterised the essence of a defensive posture as

> an attitude of alert expectation. We wait for the moment when the enemy shall expose himself to a counter-stroke, the success of which will so far cripple him as to render us relatively strong enough to pass to the offensive ourselves.
>
> (ibid.: 32)

In Corbett's (ibid.: 39) view, the defensive, 'provided we preserve the aggressive spirit, could enable an inferior force to strike a winning blow in the enemy's unguarded moments'.

Corbett's defence of the defensive is remarkably similar to Mao Zedong's exposition of the strategic defensive in his 1936 'Problems of Strategy in China's Revolutionary War'. Defining the primary problem of the Red Army following the Long March as one of conserving its strength while awaiting an opportunity to defeat the enemy, Mao advocated a strategy of 'active defence', which he described as 'offensive defence, or defence through decisive engagements'. He contrasted 'active defence' with 'passive defence', also known as 'defensive defence' or 'pure defence'. In Mao's opinion (1963:

103), 'passive defence is actually a spurious kind of defence, and the only real defence is active defence, defence for the purpose of counter-attacking and taking the offensive.'

For both Mao and Corbett, skilful exploitation of geography or terrain is a key requirement for a successful defensive strategy. Mao noted that

> one advantage of operating on interior lines is that it makes it possible for the retreating army to choose terrain favourable to itself and force the attacking army to fight on its terms. In order to defeat a strong army, a weak army must carefully choose favourable terrain as a battleground.
>
> (Mao 1963: 103)

Corbett (1911: 73), approving Moltke's doctrine that the strongest form of war is that of the strategic offensive combined with the tactical defensive, noted that:

> the use of this form of war presupposes that we are able by superior readiness or mobility or by being more conveniently situated to establish ourselves in the territorial object before our opponent can gather strength to prevent us. This done, we have the initiative, and the enemy being unable by hypothesis to attack us at home, must conform to our opening by endeavouring to turn us out. We are in a position to meet his attack on ground of our own choice and to avail ourselves of such opportunities of counter-attack as his distant and therefore exhausting offensive movements are likely to offer.

Jacob Børresen (1994: 149) has coined the expression 'coastal power' to describe the advantages of the defence elucidated by Corbett. Børresen defines 'coastal power' as the sea power of the coastal state, which he in turn defines as a state

> which does not have the resources, or has chosen not to use resources, to maintain a blue water navy with a capacity to establish sea control on the open ocean, beyond the reach of its shore-based aviation and surface-to-surface missile systems.

A coastal state 'cannot therefore challenge or compete with Naval Powers on the high seas'. He notes (ibid.: 149–50):

> Confronted with 'definitive' or 'purposeful' use of force by a Naval Power, the Coastal State can normally do no more than put up a defence in coastal waters. In these waters the Navy of the Coastal

State may find natural shelter, or it can operate with the support of land-based fighter aircraft, behind minefields or under the shelter of shore batteries. The exception is the submarine. A Coastal State that operates modern submarines may represent a serious threat to the units of the Naval Power both in coastal waters and on the open ocean.

The operations of the PLA Navy's submarine fleet would not necessarily be confined to the narrow seas within the first island chain: increasingly PLA Navy submarines are training in blue water, and, according to Western military intelligence sources, in 1999 submarines from the East Sea Fleet and South Sea Fleet were undertaking much longer sea patrols around the eastern coast of Taiwan for the first time (previously only Northern Fleet submarines had undertaken such patrols) (Shambaugh 2002: 102). On 12 November 2003, a *Ming*-class submarine was reported to have been spotted by a Japanese Marine Self-Defence P-3C heading west on the surface of international waters 40 kilometres east of Satamisaki, a port town of Kagoshima Prefecture on Kyushu Island. The submarine sailed through the Osumi Strait between Kyushu and Tanegashima islands (Tsang 2003). The fact that it was sailing westward on the surface was interpreted as deliberately communicated evidence of the ability of China's submarines to evade detection by advanced Japanese and US ASW measures and to challenge Japanese and US command of Japan's western maritime approaches. The purpose of the message was to deter US and Japanese intervention in support of Taiwan by demonstrating the combat capabilities of the Chinese submarine fleet.

Despite having the capability to operate submarines with relative impunity beyond the first island chain, Chinese operational strategists will nevertheless try to exploit to the maximum the natural advantages which coastal states enjoy in their coastal environments over naval powers whose forces have been structured, trained and equipped for general command of the oceans. In such coastal environments, according to Børresen (1994: 150), defence is a relatively stronger form of combat than in the open ocean. Operations in familiar coastal, inshore and restricted waters enable coastal navies to exploit to the maximum opportunities for deception, cover and protection. Their forces are more intimately acquainted with local conditions and are trained and equipped to operate optimally in them. Recent research undertaken by Harbin University for the PLA Navy indicates that undersea mapping is a particular priority. Goldstein and Murray (2004: 178) report that Chinese naval survey units have recently produced a three-dimensional digital chart of China's coastal waters. In the absence of reliable satellite-based precision navigation capabilities, the most accurate method of determining a submarine's precise position is by bottom-contour navigation, using a secure echo-sounder whose transmissions are practically undetectable. This method of navigation is possible only if the seabed has been precisely surveyed in advance (Moore and Compton-Hall 1987: 243).

The weapons and tactics of coastal powers are also often better adapted to operations in coastal waters than are those of blue-water navies. Consequently they have more opportunities to use surprise to better tactical effect than their naval-power adversaries. And, as Børresen (1994: 165) observes,

> the Coastal Navy should have a capacity to 'bring the war to the enemy' in terms of attacks on his naval or auxiliary units in international waters or in his own home waters. In a Coastal Navy, this is primarily the task for the submarine.

Moreover, in a confrontation between the PLA Navy and United States Navy in the narrow Yellow, East and South China Seas between the Chinese coast and the first island chain, Chinese naval forces would enjoy the advantages of operating in their own home waters along shorter and multiple lines of operation. US Navy forces, on the other hand, operating along fewer and longer lines, would be more vulnerable to attack.

On the other hand, the US Navy would be handicapped by the disadvantages of a blue-water navy operating in narrow seas. Milan Vego (1999: xv–xvi) has drawn attention to the potential problems likely to confront blue-water navies in fighting wars in narrow seas:

> There is a widely held but erroneous view that a fleet capable of defeating an adversary on the open ocean could successfully operate in narrow seas. It is true that, until relatively recently, a blue-water navy did not face serious and diverse threats to operations by its large surface combatants and submarines in, for example, the Arabian (Persian) Gulf and the Red Sea and other narrow seas of Southeast Asia, but the situation is rapidly changing because many of the smaller navies operating in these waters have acquired the capabilities to challenge the unrestricted operations of a blue-water navy such as that of the US. A typical narrow sea also poses other challenges to the operations of large naval vessels because of the unique features of its physical environment, specifically, the small size of the area involved and the correspondingly short distances, the proximity of the land, the shallowness of the water and the presence of a large number of islands and islets. These features, in turn, generally limit manoeuvrability, speed and the use of weapons and sensors by one's surface ships and submarines, primarily designed for employment on the open ocean. The range of threats to the survivability of large surface combatants and submarines is greater in a typical narrow sea than on the open ocean because the

weaker opponent can use land-based aircraft, small and fast surface combatants or conventional submarines, mines and even coastal anti-ship cruise missiles to contest control of a stronger navy.

This combination of advantages which accrue to a coastal navy operating in its own environment and handicaps for a naval power traditionally trained, equipped and organised primarily for open-ocean operations would create a more level playing field for the PLA Navy in a confrontation with the US Navy than a simple balance-of-forces assessment would suggest. In strategy, as in real estate, position is all-important. As Vego (ibid.: 42) points out:

> it is the possession of power plus position that constitutes an advantage over power without position, or, more instructively, equations of force are composed of power and position in varying degrees, surplus in one tending to compensate for deficiency in the other.

Vego's observation about the variables of power and position in calculating equations of force recalls Albert Wohlstetter's notion that effective power diminishes with distance – his 'loss-of-strength gradient' (Gray 1999b: 32n). Some twenty-five centuries earlier, Sun Zi had already discovered the loss-of-strength gradient and the difficulties of power-projection over distance:

> Those who arrive first at the battleground will have sufficient time to rest and prepare against the enemy. Those who arrive late at the battleground will have to rush into battle when they are already exhausted. Thus, the person adept in warfare seeks to control and manipulate his enemy instead of being controlled and manipulated.
> (Chow-Hou 2003: 132)

Brodie (1944: 200) also observes that

> belligerents are profoundly anxious to have engagements take place in areas of their own choosing, which will always be as close to their own bases and as distant from those of the enemy as possible. The disproportion of risk involved is often great enough to overcome an initial superiority on the part of the opponent.

It is a commonplace among theorists of naval warfare to emphasise the differences between maritime strategy and land-warfare strategy which derive from the differences in the physical environment in which the two forms of warfare are waged. Brodie (ibid.: 13), for example, maintained firmly that:

The sea has few of the terrestrial complexities of the land. It has nothing resembling railroads or highways; it has no mountains, forests or rivers; it has no centres of population or industry. It is only flat waste to be traversed, and so far as nature is concerned it is almost uniformly traversable throughout. True, different areas of the sea vary tremendously in strategic importance, but these variations are imposed upon the maritime areas by the lands which delimit them.

While this is arguably true of warfare on the surface of the sea, the subsurface presents a physical environment which is every bit as complex as the terrestrial environment. The propagation of sound in water is complex, its velocity varying with temperature, pressure and salinity, and therefore with depth, location, season and even time of day. Submarine commanders are able to exploit the salinity and temperature layers of the water, the acoustic properties of the environment and the topography of the seafloor to manoeuvre tactically in the same way that commanders on the land exploit the physical terrain. And there would be few undersea environments as complex as the environment in the marginal seas and littoral waters of East Asia. The acoustic qualities of the Taiwan Strait in particular make these waters 'a nightmare for ASW operators' (Shlapak *et al.* 1999: 22). Moreover, the physical characteristics of the marginal seas directly adjacent to the Chinese coast are more likely to favour the operations of smaller diesel-electric tactical submarines of the PLA Navy than those of the larger nuclear-powered tactical submarines of the US Navy.

The seas surrounding the Chinese mainland out to the edge of the continental shelf are relatively shallow. The seafloor of the continental shelf is composed predominantly of sands, silts and mud introduced by such rivers as the Yangtze and the Min Kiang. Rock outcroppings are exposed in the Taiwan Strait because of the strong current. The mean depth of the East China Sea is 349 metres. With the exception of the Okinawa Trough, which runs from Kyushu to Taiwan along the western edge of the Ryukyu island chain (Groves and Hunt 1980: 107), the East China Sea lies on the continental shelf which slopes gently eastwards to a depth of 200 metres along the edge of the shelf, before dropping abruptly into the 2,000 metre-deep Okinawa Trough. The tide ranges from 5 metres around Taiwan to 11 metres in Hangchow Bay southwest of Shanghai. A small branch of the warm Kuroshio Current flows into the sea around Taiwan and moves northwards. The Yellow Sea, which is only 60–80 metres deep (Leier 2001: 162), is even shallower than the East China Sea. It derives its name from the sediment brought to it from the great rivers of China. In the northern part of the South China Sea, including the Gulf of Tonkin and the continental shelf surrounding the island of Hainan, the average depth is also no more than 40 metres. The island of Taiwan sits on the edge of the continental shelf. On its eastern coast, the seafloor drops sharply away into the deep ocean of the Philippine Sea. The seas off the coast of China are notorious for their violent storms and typhoons. The typhoon season runs from May to December, with the maximum frequency occurring between July and October. Taiwan is situated in an area where gales of force 7 and above occur during 20 per cent of the year (Couper 1983: 160). And in the spring, dense fogs often prevent movement of shipping in the Strait.

Diesel-electric submarines, such as the Chinese *Kilos* and *Songs*, have a particular advantage over nuclear attack submarines in shallow littoral waters like those off the China coast. Although nuclear-powered attack submarines can usually outperform conventionally powered submarines, particularly in terms of speed and endurance, they perform less than optimally in shallow littoral waters. Since ASW detection, targeting and attack still rely predominantly on acoustic methods, diesel-electric submarines running on electric motors alone – perhaps 90 per cent of the time on patrol – have the great advantage of being able to operate in virtual silence over the full band of sonic frequencies (Moore and Compton-Hall 1987: 30). Diesel-electric submarines are able to use the undersea topography to evade ASW sonar detection by settling on a shallow (less than 300 metres) seafloor, switching off their engines and closing their seawater inlets. Nuclear submarines, on the other hand, cannot turn off all onboard machinery if they are to operate successfully, and are therefore susceptible to passive detection from the steady signature emitted by this machinery. Reactor main coolant pumps are a particular source of noise in nuclear-powered submarines (Moore and Compton-Hall 1987: 42). Moreover, nuclear submarines cannot sit on the bottom, particularly a muddy bottom such as that of the Taiwan Strait, without clogging vital inlets to condensers. In such an environment, a bottomed diesel-electric submarine is relatively safe from homing weapons which use active sonars to detect targets. Doppler seekers are ineffective in detecting a motionless submarine which, by definition, emits no Doppler signature (Friedman 1995: 55). In addition, unlike nuclear-powered submarines, diesel-electric boats are able to take advantage of waters such as those off the coast of China constantly traversed by numerous diesel-powered freighters and trawlers, to disguise their acoustic signatures. According to Vego (1999: 37),

> man-made noise, caused by ship traffic in harbours and their approaches, by busy shipping lanes, and by coastal settlement activities, is, on average, 5–10 decibels (dB) higher in shallow waters close to shore than it is in deep waters, especially at frequencies greater than 500 Hertz (Hz).

In the narrow and congested seas off the China coast, notorious for its sudden and violent storms, the seafloor is likely to be littered with sunken ships among which submarines can hide to confuse sonar operators and magnetic anomaly detection devices. The roughness of the sea surface in shallow waters is also more significant in determining the performance of sonar than it is in the open ocean, where the vertical temperature structure of the water is a more significant factor (Vego 1999: 35).

Submarines can also take advantage of the fact that in the tropical and subtropical waters of the East and South China Seas (Taiwan straddles the Tropic of Cancer), a surface layer of warm water can trap the signals of hull-mounted sonars, reducing their capability to negligible range against a submarine at even modest depth (Friedman 1995: 55). Seas such as those off the China coast into which a number of major rivers flow, depositing silt on the seafloor and lowering the saline content of the seawater, also create problems for sonar detection when the shallow salt water mixes with fresh water and creates varying layers of salinity. These layers reflect or refract the sonar beams (ibid.: 56).

The use of non-acoustic ASW techniques is a potential wild card in assessing the comparative strengths in submarine and anti-submarine warfare of the PLA Navy and the US Navy in Chinese littoral waters. Some non-acoustic techniques, such as high frequency direction-finding (HF/DF), and magnetic anomaly detection, are well known. But other techniques, which undoubtedly played an important role during the Cold War, remain highly classified because of the continued relevance of both the methods and sources used (Cote 2003: 90).

Both the Soviet Union and the United States are known to have investigated various techniques for remote, space-based detection of submarines using non-acoustic methods (Friedman 2000: 201). Since the 1960s, the US Navy has been investigating the use of blue-green lasers as a means of detecting submarines down to a depth of around 150 metres. In the early 1990s, the US Navy was also conducting research involving a wake detector, a high-powered laser and the detection of bioluminescence (ibid.: n. 39, 350). Another method investigated by the US Navy is the detection of the 'Bernouilli hump', a slight anomaly in the surface height of the sea caused by submerged objects, including submarines, which would in theory be detectable from space (ibid.: 202). Submerged submarines also produce a characteristic wake – known as a Kelvin wake – on the surface of the sea which may be detectable against the random movement of the sea. In 1999, it was alleged that in 1997 an American physicist working at Lawrence Livermore Laboratory had given details of a US submarine wake detection project to the Chinese. The project reportedly involved the use of synthetic aperture radar and supercomputers to distinguish faint submarine wakes from surrounding waves (ibid.). Bioluminescence – caused by the reaction of plankton to the disturbance of passing submarines – may also provide a means of detecting the presence of submarines.

If the US Navy has achieved any success in pursuing these methods of non-acoustic detection of submarines, it could alter the balance of advantage in submarine and anti-submarine warfare in East Asian seas. But Friedman's (2000: 205) observation that 'continued heavy investment both in submarines and sonars, both in the United States and abroad, suggests that no one has been able to achieve reliable non-acoustic detection from the air, let alone from space' perhaps best encapsulates the state of progress of these avant-garde techniques in submarine detection.

7

DISPUTING US COMMAND OF
THE CHINA SEAS

Sea denial and anti-access strategies

As General Beaufre points out, strategy implies not only that armed force is organised and used, but that this is done with the purpose of attaining polit- ically defined goals, and in the expectation that the attainment of those goals will be resisted by an opposing force. Thus, for Beaufre (1963: 45), strategy is 'a dialectic of forces or more exactly the art of the dialectic of wills using force to resolve their conflict'. More succinctly, Clausewitz (1976: 83) considered that 'war is nothing but a duel on a larger scale'.

The PLA Navy's decision to enhance its undersea warfare capability is thus also to be seen as a dialectical response to the post-Cold War evolution of the maritime strategy of its main potential adversary, the United States. The US Navy and Marine Corps' new maritime strategy was first articulated in a White Paper published in 1992 entitled *From the Sea: Preparing the Naval Service for the 21st Century*. The concepts in this paper were refined in a 1994 supplement entitled *Forward ... From the Sea* and the operational aspects of the new maritime strategy were elaborated in two documents published in 1997, 'The Navy Operational Concept' and the Marine Corps 'Operational Maneuver from the Sea Concept' (Weeks and Meconis 1999: 126). Together, these documents spell out 'a *fundamental* shift away from open-ocean warfighting *on* the sea toward joint operations conducted *from* the sea' (ibid.: 127). The central concept of this new maritime strategy is to provide 'Naval Expeditionary Forces', designed for joint operations, to project force by 'operating forward from the sea'.

The 1991 and 2003 wars against Iraq – as well as the smaller-scale Operation Desert Strike in 1996 and Operation Desert Fox in 1998, Operation Allied Force against Serbia and Serbian forces in Kosovo in 1999, and Operation Enduring Freedom against Taliban and al-Qaeda forces in Afghanistan in 2001 – have all provided Chinese strategic analysts and planners with graphic, practical demonstrations of joint operations conducted, at least in part, from the sea. Memories of the sustained naval air campaigns conducted by the United States in Korea and Vietnam are no doubt also prominent in the minds of Chinese strategic planners.

The Chinese submarine fleet, spearheaded by the *Kilo-* and *Song*-class boats, provides Beijing with its most effective sea-denial instrument, well suited to preventing the US Navy from approaching the Chinese littoral, or at least depriving it of its freedom of action in the zone which extends at least up to 200 nautical miles from the Chinese coast. In any crisis or conflict involving the United States and its allies, the PLA Navy's submarine fleet would play a leading role in an anti-access strategy designed to keep United States naval forces away from the Chinese coast.

The importance of the role of the PLA Navy's submarine fleet in this anti-access strategy would be all the greater for the fact that the other components of Chinese sea power which could theoretically contribute to this strategy, particularly the PLA Air Force, the PLA Navy Air Force and the PLA Navy surface units, would have difficulty competing with even second-rank regional military forces, let alone a first-class maritime power such as the United States. Currently, only China's two *Sovremenny* destroyers with their systems of eight SSN-22, 'Sunburn' (designated 'Moskit' by the Russians) anti-ship missiles with a range of 180 kilometres and a speed of Mach 2.5, pose a real threat to American surface combatants and submarines. A Chinese *Sovremenny*-class destroyer first flight-tested one of these missiles in September 2001 (Gertz 2001). But even these weapon systems lack sufficient stand-off range to challenge US offensive forces (Ross 2002: 67). Moreover, the PLA Navy surface units – no doubt because, like their Soviet predecessors, they were designed to operate within range of land-based air protection – are very vulnerable to air attack. They are equipped with short-range HQ-7 and HQ-61 air defence systems which provide them with only a restricted defensive zone (Kaplan 1999). As Robert Ross (2002: 67) comments, 'US capabilities would be even more effective in targeting Chinese surface assets at sea than they have been in targeting enemy assets in deserts, such as in the Gulf War and the war in Afghanistan.'

Beijing is nonetheless moving to make good these deficiencies. The PLA Air Force has taken delivery of some 40 SU-30MKK fighter bombers and almost 100 SU-27 fighters. These aircraft will be equipped with a variety of high-performance weapons, including the Russian long-range AA-12 Adder. China's Air Force is also increasing its numbers of aerial refuelling tankers and airborne surveillance aircraft (Goure 2003). Twenty-four SU-30MK2s, delivered in February 2004, have been designed especially for the PLA Navy. These aircraft will give the PLA Navy the equivalent air strike power of a US aircraft carrier battle group (Kanwa 2003a). But until these new systems become fully operational, submarines remain one of Beijing's few real means for preventing the United States, with its precision-guided, long-range weapons, from waging war safely beyond the range of Chinese forces.

The PLA Navy's emphasis on the modernisation and expansion of its submarine arm is a dialectical response not only to the US Navy's post-Cold War emphasis on developing the capabilities and doctrine to enable it to counter anti-access strategies and to engage more effectively in littoral warfare, but also to the US military forces' transformation strategy. The goal of this transformation strategy, outlined in Chairman of the Joint Chiefs of Staff document *Joint Vision 2020* published in June 2000, is to develop the capabilities, organisational structures, training and doctrine which will enable US military forces to be dominant 'across the full spectrum of military operations' (USDoD 2000: 1). The foundation for the achievement of this overarching objective of 'full spectrum dominance' is information superiority – 'a key enabler of the transformation of the operational capabilities of the joint force and the evolution of joint command and control' (ibid.: 3). The aim of information superiority is to achieve an almost complete reading of the deployment of enemy forces. By 2005, US military forces hope to be able to detect around 90 per cent of military targets within a 320 square-kilometre area (Richardot 2002: 229). The US Navy plans to achieve information superiority through the concept of 'Network Centric Operations', the stated aim of which is to:

> [promote] superior knowledge that will foster a shared, near real time understanding of the battlespace, complementing the Navy's command of the seas with speed of command. The result will be battlespace dominance, operating inside the sensor and engagement timeline of an adversary to foreclose the effectiveness of anti-access strategies.
>
> (US Navy 2002)

Joint Vision 2020 predicts that in response to the evolution of US military capabilities, potential adversaries of US military forces will adopt niche capabilities and asymmetric approaches in an attempt 'to avoid US strengths and exploit US vulnerabilities ... to delay, deter or counter US military capabilities' (USDoD 2000: 4). China's acquisition and development of advanced submarine warfare technology is an instance of Beaufre's dialectical process in operation: at least one potential adversary is anticipating and adapting to the ongoing transformation of US military capabilities by developing a capability which, for the time being at least, is relatively secure from 'information dominance' of US forces.

Searching for America's Achilles' heel: submarines versus aircraft carriers

The effectiveness of China's undersea warfare capability as a deterrent to intervention by the United States in a Taiwan conflict depends on the credibility of the threat which it poses to US forces deployed in the region. Many Western analysts have suggested that Chinese strategists believe that the loss of an aircraft carrier, or several other United States Navy ships, with several thousand casualties, would be worth more to Washington than the 'loss' of Taiwan (e.g. Burles and Shulsky 2000). In the eight years following the 1996 Taiwan Strait crisis the PLA is reported to have conducted seven major exercises, codenamed 'Project 968', simulating the sinking of aircraft carriers (Ching 2004). According to Mark Stokes (1999: 82), the China Aerospace Corporation's (CASC) Third Academy, China's most important centre for research and development of cruise missiles, has 'conducted in-depth targeting assessments of US aircraft carriers, including fleet defense assets, characteristics of radar and infrared emissions, and electronic combat capabilities'. The Third Academy is also working on the development of a capability to launch cruise missiles from submerged platforms (Stokes 1999: 85). The new imported *Kilo*-class submarines will be capable of launching cruise missiles while submerged, and the indigenous *Song*-class boats are designed to carry the YJ-82, China's first anti-ship cruise missile capable of being launched from a submerged submarine.

The PLA Navy's most effective conventional weapon against a US aircraft carrier, however, is likely to be the *Kilo*-class submarine's wake-homing torpedoes. Large, armoured warships are inherently difficult to sink or disable with hits above the waterline, unless the missiles manage to penetrate a vital area of the ship such as its magazine or combat information centre. Indeed, as Friedman (2001: 251) observes, 'hits above the waterline (i.e. most missile hits) may *never* directly sink a large ship'. Multiple hits could nevertheless cause significant damage and casualties. And if an attacker were lucky enough to damage aircraft concentrated on the flight deck or to severely damage the flight deck itself, it would render the aircraft carrier largely ineffective. Underwater weapons, however, 'are inherently far more lethal than their above-water counterparts, because they can flood and thus sink a ship' (ibid.: 253). Torpedoes are also generally less susceptible to countermeasures than missiles (Moore and Compton-Hall 1987: 23). Even so, with its large hull and strong, armoured sides designed to counter earlier contact torpedoes – as opposed to modern torpedoes which explode under a ship rather than in her side – a current US aircraft carrier would have a good chance of surviving a hit even by the heaviest (non-nuclear) torpedoes in the Russian or Chinese arsenals (Friedman 2001: 251). The Soviet Navy used to calculate that effective action against a US nuclear aircraft carrier would require at least 25 torpedoes and 15 cruise missiles (Weir and Boyne 2003: 223). The inherent strength of modern US aircraft carriers was demonstrated in 1969 when the USS *Enterprise* suffered a catastrophic accident in which nine of its 500 pound bombs detonated, with the explosive power roughly equivalent to half a dozen Russian cruise missiles. Despite 27 deaths and 300 injuries among the crew members, the *Enterprise* resumed strike operations within hours (Thompson 2001: 27). This implies that for an effective attack against a US aircraft carrier, the PLA Navy would have to orchestrate a near simultaneous attack by a number of platforms, including submarines, surface combatants, aircraft and possibly land-based missiles. Such an attack would in turn require near real-time sources of information and imply a high level of command and control to coordinate the combined action of different services and branches of services.

Provided that the *Kilo* submarines could get close enough to an American aircraft carrier to fire their torpedoes, they could nevertheless pose a serious threat to these vessels. As Paul H. B. Godwin (2003b: 43) has noted:

> The most effective offensive capability being acquired by the PLAN is the *Kilo* diesel-electric submarine acquired from Russia, especially the type 636 ... Even considering the ASW capabilities found in a US CVBG, this submarine presents a very real threat. Assuming the next generation Chinese SSN (the 09–3 program) benefits from Russian technology, weaponry, and design assistance, the submarine threat to US naval operations over the coming decade is going to increase significantly.

The potential lethality of advanced conventional diesel-electric submarines against US Navy carriers and their escorts was demonstrated in a September 2003 exercise which pitted Australia's *Collins*-class submarines against American attack submarines and Singaporean ASW surface vessels in the Indian Ocean. At certain speeds the *Collins*-class submarines, like *Kilo*-class submarines, can render themselves virtually undetectable, becoming 'underwater black holes'. During the exercise, one *Collins* submarine 'sank' a Singaporean surface unit fitted with some of the world's most advanced ASW equipment, while another 'sank' an American nuclear attack submarine (McPhedran 2003: 11). In early October 2003, the Indian and US navies conducted joint exercises in the Arabian Sea during which they recorded the electronic and acoustic signatures of each other's submarines. But the Americans were reported to have been disappointed that Russia did not allow India's *Kilo*-class submarines to participate in the exercise (Raghuvanshi 2003: 15). This suggests that US Navy ASW operators may not be confident of being able reliably to identify Chinese *Kilo*-class submarines in Northeast Asian seas. To improve the skills of its ASW operators against quiet diesel-electric submarines in shallow coastal waters the US Navy has reportedly asked the Swedish government to lend the US Pacific Fleet a *Gotland*-class submarine and crew to be based at the Fleet Anti-Submarine War Command in San Diego (Cavas 2004b: 1).

There is little doubt that in choosing to wager a significant portion of its limited resources on its submarine fleet, PLA strategists believe they have identified a weak point in the US Navy's otherwise formidable naval forces. In his analysis of Chinese strategists' views on the future security environment, Michael Pillsbury (2000: 70) has found that most Chinese analysts are confident that the American armed forces, although the most technologically advanced in the world, nonetheless suffer from fatal weaknesses which could allow technologically inferior Chinese forces to prevail over those of the United States. They commonly cite a 30 January 1994 *Defense News* report of a war game between the Chinese military and the US Navy in the Pacific at the US Naval War College in which Chinese forces defeated US forces. They frequently refer to Chinese 'defeats' of American forces in Korea and Vietnam. Chinese analysts see American logistics as a particular Achilles' heel for US forces, pointing out that the United States must first cross the Pacific Ocean to wage a war in Asia, creating vulnerable lines of supply. American logistics problems are compounded by the high consumption rates of munitions and other supplies demanded by the American style of warfare. Chinese strategists also regard the United States' dependence on allies as a potential weak point, because of the problems inherent in alliances which US adversaries could exploit by dividing and disintegrating the alliance politically (Pillsbury 2000: 74). Pillsbury (ibid.: 79) quotes former Chief of Staff of the PLA, General Su Yu, as stating that one way to defeat US air force and naval air power is to strike American-controlled airbases. An article carried in a Chinese military journal has reportedly argued that 'the most effective approach of strike at time of war is to eliminate the enemy's warships right in the enemy's harbouring port' (Kanwa 2003a).

In their consideration of strategies for the weak to defeat the strong, Chinese analysts are particularly interested in the vulnerabilities of US Navy aircraft carriers and their escorts. According to their analysis, aircraft carrier battle groups are difficult to conceal because they emit numerous radar reflections and infrared and electromagnetic signatures; their flexibility is limited in shallow or littoral waters and narrow seas; their self-defence capabilities decline at night and in bad weather conditions; their anti-submarine and anti-mine capabilities are relatively poor (during World War II nine aircraft carriers – 36 per cent of the total number of aircraft carriers that sank – were sunk by submarines); and their operational effectiveness is dependent on supply ships which are vulnerable to enemy attack (Pillsbury 2000: 84). For PLA strategists, therefore, a modernised and potent submarine force acquires the characteristics of a strategic weapon, enabling Chinese forces to strike the military centre of gravity of their more powerful opponents in conformity with Sun Zi's advice in dealing with a more numerous and disciplined enemy who is about to advance: 'be first to capture something that the enemy treasures most and he will accede to your demands' (Chow-Hou 2003: 326).

Command, control, communications, computers, intelligence, surveillance and reconnaissance (C⁴ISR)

As Corbett (1911: 159) observed, one of the principal distinguishing characteristics of naval warfare, as opposed to warfare on land where it is possible to determine with some precision the limits and direction of the enemy's movements, is that:

> afloat neither roads nor obstacles exist. There is nothing of the kind on the face of the sea to assist us in locating him and determining his movements … and there is practically nothing to limit the freedom of his movement except the exigencies of fuel. Consequently in seeking to strike our enemy the liability to miss him is much greater at sea than on land, and the chances of being eluded by the enemy whom we are seeking to bring to battle become so serious a check upon our offensive action as to compel us to handle the maxim of 'Seeking out the enemy's fleet' with caution.

Seeking out the US Navy's aircraft carriers is thus the first task that the PLA Navy would have to accomplish before attempting an attack. In other words, to disable or sink a US Navy aircraft carrier, PLA forces must first find one, positively identify it among the several surface ships which make up a carrier strike group (typically the aircraft carrier itself, and at least a guided missile cruiser, a couple of guided missile destroyers and a combined ammunition/oiler/supply ship) (US Navy 2004), and then track it as it moves across the sea at a top speed of some 30 knots. Despite being the world's largest warships (*Nimitz*-class carriers are 317 metres in length, displace around 87,300 metric tonnes and have flight decks 76 metres wide) (Weeks and Meconis 1999: 142), locating, identifying and tracking US aircraft carriers would not necessarily be an easy task for the PLA. With its existing air- and ground-based reconnaissance capabilities, the PLA currently only has the ability to monitor an area of up to some 200 nautical miles within line of sight of its borders (Stokes 2000: 112). At the outset of a conflict, US aircraft carriers would no doubt operate outside this 200 nautical mile radius of the Chinese coast, relying on the 600 nautical mile unrefuelled combat radius of its strike aircraft to give it the reach it needed to attack coastal targets (Thompson 2001: 13). Moreover, if, despite the limited range of China's ocean surveillance systems, the PLA Navy succeeded in locating and tracking a US Navy aircraft carrier, the information on the identity, position, speed and course of the ship would then have to be conveyed to command centres on land and in onboard combat direction centres for processing, evaluation and decision. Strike units would have to be manoeuvred to a position to bring the target into the effective range of their organic sensors and weapons, and firing solutions worked out for an attack.

So while in theory its *Kilo* submarines may well represent one of the PLA Navy's most promising means of attacking a US aircraft carrier, in practice a successful attack would imply that the PLA Navy had succeeded in overcoming not only the problem of locating, identifying and tracking the potential target but also two significant limitations that have characterised the use of the submarine against enemy surface vessels from the start: the inherent limitations of its organic sensors to locate its targets and its low speed relative to that of most surface combatants. Brodie noted in 1944 (71) that the submarine's 'slow speed, especially when submerged, often prevented it from launching an effective attack'. It was this factor which largely dictated submarine tactics during the First World War. For example, British E-class boats, the first truly formidable and reliable submarines, were capable of a surface speed of 16 knots and a submerged speed of 10 knots (Miller and Jordan 1987: 26). The submerged speed of World War I submarines was therefore only about half that of the best speed of dread-nought squadrons of around 20 knots (Keegan 1993: 122). The relative speeds of conventionally powered submarines and major surface combatants have remained almost unchanged: a *Kilo*-class submarine is capable of a submerged speed of 17 knots, compared with the 30 knots of a *Nimitz*-class aircraft carrier. In the First World War, as a consequence of the slow speed of submarines relative to that of the enemy surface combatants, submarines, in Keegan's (1993: 216) words,

> were confined to operating against the approaches to naval ports and in what naval strategists call 'pelagic' areas – confined waters in frequent use by the enemy where his movements obeyed laws of probability, such as the Baltic, the Black Sea, Channel, North Sea and parts of the Mediterranean.

In China's case, even her *Han*-class (Type-091) SSNs, with a submerged speed of around 25 knots, would have difficulty in keeping up with a US carrier group. In any case, outdated *Han*-class submarines would have difficulty evading US ASW measures. This means that the PLA Navy's most promising course would be to deploy its quiet *Kilo*- and *Song*-class diesel SSKs to stake out the choke points between the chain of islands along the edge of the East Asian continental shelf and lie in wait for carrier battle groups as they make their way into the semi-enclosed seas off the China coast – much as Admiral Scheer positioned his boats along the probable routes of the Grand Fleet in the North Sea during the First World War. But to do this would mean that Chinese submarines would have to run the gauntlet of the ASW measures operating from the bases in the Japanese islands and from the carrier group itself.

Thus, given the current level of development of the their C⁴ISR capabilities, PLA forces would need favourable circumstances – and a good deal of luck – to be able to find, track and target a US Navy aircraft carrier successfully even within this 200-nautical-mile perimeter, let alone beyond. In its 2004 report to Congress on the PRC's military capabilities (USDoD 2004b: 24), the Pentagon observed that 'China's development and deployment of state-of-the-art ISR capabilities are uneven and will further complicate the

PLA's ability to train in a realistic joint warfare environment and ultimately fight in a modern battle. Currently targeting is a problem.' The Pentagon's report goes on to note, however, that with China's emphasis on improving its space-based imagery and reconnaissance satellites 'this likely will improve over the next decade'.

The PLA is indeed well aware that the answer to this problem of finding, identifying, tracking and targeting US aircraft carriers is most likely to be found in space. Indeed, according to Dean Cheng (2003: 27), 'Chinese military writers clearly recognise that space operations will be a fundamental part of future military operations.' The logic of the PLA's recourse to space-based ISR assets as a solution to the problem of finding and attacking US aircraft carriers is similar to that of Soviet strategic planners in the late 1950s. In the 1950s and 1960s, the Soviet Navy's greatest concern was to develop ways to combat US aircraft carrier strike groups which could attack targets on shore at ranges of hundreds of kilometres. This threat assumed a vital importance for the Soviet high command and strategic planners once US carrier aircraft became capable of delivering nuclear weapons to cities deep within the Soviet homeland. It thus became essential for Soviet military planners, in the words of Marshal V. D. Sokolovskiy, 'to destroy the attack carriers before they launch their aircraft' (Ranft and Till 1983: 155). The initial Soviet response to this threat was to develop a range of land-based coastal defence missiles and then, as the range of the carrier-borne strike aircraft increased, a variety of long-range anti-ship missiles for delivery by surface vessels, aircraft and submarines. Maritime reconnaissance aircraft and land-based high frequency radio direction-finding (HF/DF) sensors were used to locate the carrier groups and to cue strike aircraft, ships and submarines to attack their targets. Before long, however, the reconnaissance aircraft, as well as the surface combatants and strike aircraft became increasingly vulnerable to Western carrier-borne fighters as the carrier groups began to operate beyond the range of Soviet land-based air protection (ibid.: 156). At the same time, the land-based HF/DF system became less reliable as a means of finding targets the further out to sea the carrier groups operated. Moreover, in the 1960s the US Navy began to use satellites rather than HF radio for its long-range communications, further reducing the value of HF/DF sensors as a means of locating its carrier groups (Friedman 2000: 149).

The solution to the Soviet Navy's problem of locating, tracking and targeting US aircraft carriers was a space-based reconnaissance and detection system, first proposed in 1959–60 (ibid.: 155). The proposal was adopted in 1961, and a pair of Soviet ocean reconnaissance satellites, one for conducting active radar reconnaissance and the other for passive, electronic surveillance measures (ESM), entered into service in the early 1970s. These were known in the West respectively as RORSAT (Radar Ocean Reconnaissance Satellite) and EORSAT (ELINT Ocean Reconnaissance

Satellite) (ibid.: 157). The CIA estimated that RORSAT was capable of detecting cruisers and destroyers under favourable conditions and large ships such as carriers under most conditions (ibid.: 158). Western analysts estimated that EORSAT was capable of locating targets within about 2km (ibid.: 161). The whole system of paired active and passive satellites, control stations, downlinks and surface stations was collectively called the Legenda system (ibid.: 162).

Although satellite ocean reconnaissance was a major breakthrough for the Soviet Navy, it soon realised, however, that simply locating, identifying and tracking prospective targets was not enough. To make use of the information provided by the Legenda system, it had to be processed and communicated to deployed units which had to be cued into position to attack. This meant that complementary means of long-range communication and precise navigation had to be developed. Again, the best solution was to be found in space. In 1962, therefore, Moscow commissioned a communications-navigational satellite known as *Tsiklon*, together with a complementary shipboard receiver known as Tsunami. The system entered into service in 1971 (ibid.: 163).

China is clearly following the same path as both the Soviet Union and the United States in developing its space-based C^{4}ISR capabilities. With respect to the reconnaissance and surveillance elements, the Pentagon (USDoD 2003: 43) reports that: 'China is placing major emphasis on improving space-based reconnaissance and surveillance, including electro-optical, synthetic-aperture radar, and other satellite reconnaissance systems. These systems, when fully deployed, are expected to provide a regional, and potentially hemispheric, continuous surveillance capability.'

China's first recoverable reconnaissance satellite, the *Fanhui Shi Weixing* (FSW) was launched into orbit in 1975, making the PRC at the time one of only three space-based reconnaissance powers. The Pentagon (ibid.: 32) also reported in 2003 that China had launched its first oceanological satellite – with the purpose, according to the Chinese press, of collecting precise data about the ocean's colour and temperature – on 15 May 2002. According to American analyst Mark Stokes (2000: 113), CASC is developing at least four space-based systems which will expand the PLA's ability to support strike operations further from Chinese shores. According to Stokes (ibid.), before the end of the current decade, China's space-based surveillance architecture could have at least four components:

1   synthetic aperture radar (SAR) satellites for all weather, day/night monitoring of military activities;
2   electronic reconnaissance satellites to detect electronic emissions in the western Pacific;
3   mid-high resolution electro-optical satellites for early warning, targeting, and mission planning; and

4   a new generation of high resolution recoverable satellites for intelligence analysis.

Although a central purpose of China's evolving space-based ISR assets will be to provide precise targeting data for the PLA's large and increasingly lethal arsenal of theatre-range ballistic and cruise missiles, Stokes notes that 'according to Chinese sources, SAR and electronic reconnaissance satellites would serve as important components of an ocean monitoring network for detecting and tracking naval activities, to include carrier battle groups and submarines'.

China's first-generation SAR satellite, the *Haiyang-1* (HY-1), an oceanographic research satellite, was launched in 2002. According to the Pentagon (USDoD 2003: 43), at least two additional satellites in this series, HY-2 and 3, are expected by 2010. This series of satellites is also expected to support earth observation, communications and navigation functions. PLA strategists have noted that the advantage of space-based SAR systems is that they can see through clouds, rain and fog in order to detect and track ships and submarines in shallow water. Research has already begun on a second generation of more sophisticated SAR satellites (Stokes 2000: 114).

The Pentagon (USDoD 2003: 41) has also reported that China is interested in the second component essential of any ocean-monitoring network, electronic intelligence (ELINT) or signals intelligence (SIGINT) reconnaissance satellites, and that Chinese scientists may be developing a system of data relay satellites to support global coverage. ELINT satellites monitor radars and other sources of electromagnetic-spectrum emissions to pinpoint the precise location of their transmitters, and to determine their purpose and operating characteristics. Provided its receiver is precise enough and that other Chinese ELINT platforms have collected sufficient information on the electronic signatures of US Navy units, a space-based ELINT reconnaissance capability would enable the PLA Navy not only to distinguish carrier formations but to identify individual ships within the formation. According to a Russian article published in 1993 (Andronov), the United States' space-borne ELINT system, Whitecloud, is the principal means of over-the-horizon (OTH) reconnaissance and target designation for the US Navy's weapons systems.

According to the Pentagon (US Senate 2004), China is also developing small satellites and micro-satellites – weighing less than 100 kilograms – for missions that include remote sensing and networks of electro-optical and radar satellites. In 1998, CASC announced its intention to field a tactical small satellite imaging constellation and associated ground receiving stations. The system will include four electro-optical (EO) and two SAR satellites.

In addition to the PRC's drive to improve its space-based C^{4}ISR capabilities, the Pentagon (USDoD 2003: 43) also reports that 'AEW aircraft, long-range UAVs, and over-the-horizon radar will enhance its ability to detect, monitor, and target naval activity in the western Pacific Ocean'. Naval ISR programmes include the Y-8 AEW aircraft and efforts to procure or produce an AWACS which will complement existing ISR platforms such as the Tu-154 aircraft equipped for ELINT collection missions (USDoD 2003: 44). The Pentagon (ibid.) also assesses that China may have as many as three OTH sky-wave radar systems to track maritime movements in China's contiguous seas 'and most likely to serve as part of an effort to develop the capability to track and target US aircraft carriers'.

China also launched its first two navigational satellites in 2000. The Beidou Navigation System (BNS) comprises two domestically built satellites in geostationary orbit capable of providing full-time, all-weather navigational information. This makes China only the third state to deploy a satellite navigation system following the United States' GPS and the Soviet Union's GLONASS system (Cheng 2003: 38). Satellite-based navigation systems have become essential for the accurate targeting of precision weapons such as anti-ship and land-attack cruise missiles and surface-to-surface ballistic missiles. And because of a submarine's inherently limited vision and speed, the ability to navigate precisely would be crucial for any attempt by PLA Navy submarines to intercept US aircraft carriers. Precise navigation is also important for mine warfare, since both mine layers and mine hunters need to know the exact location of minefields. Until China deployed its own satellite navigation system it had to rely on GLONASS and GPS, creating a dependence on potentially unreliable foreign systems which could become a critical vulnerability in time of war. An indigenous satellite navigation system thus plays a fundamental role in the PLA's defence modernisation programme, in its naval strategy and in its overall military strategy. Although China's first generation of navigation satellites may not provide sufficient accuracy for the precise targeting of missile systems, CASC is now designing a more complex global navigation system which could provide the necessary degree of precision (Stokes 2000: 182).

Despite these advances in China's space-based ISR programmes, the PLA still has to overcome formidable obstacles to put in place a space-based system capable of reliably locating, tracking and targeting US aircraft carriers. To put the challenge facing the PLA Navy into perspective, Loren Thompson (2001: 12) has calculated that for the South China Sea alone, an area of some 300 million square miles:

> three bands of 46 satellites each (138 spacecraft in all) operating in 40-degree inclined polar orbits would be required to provide constant monitoring. The size of the satellite constellation is driven by the need for continuous coverage and high resolution. High resolution dictates low-earth orbits. Low-earth orbits in turn dictate how many satellites must be in each band to avoid gaps in coverage,

and also how many bands there must be to cover the whole sea given a 300 nm field of view per band. Continuous coverage could be achieved from higher altitudes using fewer satellites, but resolution would deteriorate to a point where it was no longer suitable for use as targeting data.

In addition to surveillance systems for the South China Sea, China would need more satellites to cover the eastern and north-eastern approaches to its coast. It is likely, therefore, given Beijing's finite resources, and the fact that Chinese space technology is still generations behind that of the United States, that space-based capabilities that can reliably find, track and target US aircraft carriers will be beyond its reach for a long time to come.

Anti-submarine warfare (ASW)

The submarine acquires an even greater potential as a strategic weapon to Chinese naval strategists because of their belief that the US Navy's anti-submarine warfare capabilities would be challenged in the difficult conditions presented by the narrow seas off the Chinese coast. This belief is not without substance (see Chapter 6). The US Navy has acknowledged that it currently lacks a number of key war-fighting capabilities it needs for operations in littoral environs, including the detection and neutralisation of enemy submarines in shallow water, mine warfare capabilities and defending its ships against cruise missiles (USGAO 2001b: 2).

The Americans have been worried about erosion of their anti-submarine warfare capabilities since the end of the Cold War (Morgan 1998). The US Navy's ASW skills were forged through hunting Soviet nuclear submarines during the Cold War. For the most part, the hunting grounds for Soviet submarines were in blue waters, with the emphasis on tactics using passive sonar. The US Navy tended to leave ASW against diesel-electric submarines, in which active acoustics played a central role, to allied navies (Cote 2003: 38). Today, however, ASW activity has shifted towards green waters in littoral regions. ASW techniques developed for the deep ocean and countering nuclear-powered submarines are not necessarily well suited to shallow littoral waters where diesel-electric submarines pose a more significant immediate threat, where the ambient noise of coastal maritime traffic, waves and marine life are louder and more numerous, and where the sea currents, salinity and temperature variations are more unpredictable. Shallow water eliminates the deep sound channel and convergence zone propagation paths which create the optimum conditions for the passive sonar tactics which gave the US Navy the ASW edge over the Soviet Navy during the Cold War (Cote 2003: 66). The conditions commonly found in warm littoral waters make ASW a much more complex task in green waters than in the cold northern oceans. Moreover, in littoral waters and narrow seas ASW units are much more vulnerable to mine warfare and attack by shore-based defences. Asked in a recent interview whether today's submarine force could consistently locate, identify and target modern diesel-electric submarines operating in the littoral, Rear Admiral Paul Sullivan, director of the United States Navy's Submarine Warfare Division, responded:

> That's a complex problem. The oceans are not transparent. Finding any submarine anywhere is not a given. The littorals bring their own set of significant issues such as the impact of currents, the mixing of water. You have a lot more shipping and interference. It makes the acoustics both different and difficult. It's like trying to get across [Interstate] 395 in the middle of a busy rush hour versus being out in the open. The acoustic conditions make it not only hard to find [diesel-electric subs] but hard for your potential killing mechanisms, your torpedoes, to do their job.
>
> (*Seapower* 2003)

Since, as Shlapak *et al.* (1999: 44n) observe, 'virtually all advanced navies, including the US Navy, rely on their own submarines as a primary weapon against an enemy's undersea forces', the reduction in the number of attack submarines (SSN) in the United States fleet has also reduced the ASW capability of the US Navy. Since the end of the Cold War, the US Navy has sharply reduced the number of SSNs in its fleet: from 65 units in 1998, to 57 in 1999, and 50 in 2003 (Richardot 2002: 96). At the same time, the United States has virtually stopped building new attack submarines: the US Navy built only four between 1991 and 2000, compared with 28 between 1981 and 1990. Most of the current force of the US Navy's attack submarines belong to the *Los Angeles*-class (SSN-688), the first of which was built in 1976. The normal service life of these vessels is around 30 years, after which their reactors become brittle and their structural welds begin to show signs of fatigue. This means that most of the boats built during the 1980s will be retired between 2011 and 2020. Unless the United States increases its production rate of the new *Virginia*-class SSNs above the current rate of one a year, the US Navy is likely to face a serious deficit in its undersea warfare capability during the next decade (Thompson 2000). In 2003, the navy planned to increase the rate of building new *Virginia*-class boats to two a year starting in 2007, but in its 2005 budget this plan has been deferred by two years until 2009 (Ratnam 2004: 22). The US$368.2 billion defence spending bill passed by the Congressional Appropriations Committees for budget year 2004 approved a five-year plan to buy five *Virginia*-class SSNs, two short of the seven that the navy had requested (Mathews 2003: 18). As of June 2004, the United States Navy had 10 *Virginia*-class SSNs under contract, with six in various stages of construction (Butler 2004: 53). Nonetheless, American naval analysts still question whether the US Navy will have as many of these vessels as it will need. The *Virginia*-class SSN, the first of which was delivered to the US Navy in June 2004, is significantly more capable than the current *Los Angeles*-class SSNs, and is specifically designed for operations in both deep and littoral waters.

In June 2000, Rear Admiral Konetzni, commander of the Pacific Fleet's submarine forces, testified that the Pacific Fleet had 26 attack submarines, far short of the 35 needed to perform the missions associated with aircraft carrier group support and engagement with allies (FAS undated). In 1999, a study by the Chairman of the Joint Chiefs of Staff concluded that a force structure of below 55 attack boats in 2015 and 62 boats in 2025 would leave regional military commanders-in-chief with insufficient capability to respond to urgent critical demands. The study concluded that 68 SSNs in 2015 and 76 SSNs in 2025 would be required to meet other high priority but less critical demands (USGAO 2001a). Although some measures have been undertaken or are being considered to relieve the pressure on the US submarine fleet, including the forward-basing of at least three SSNs at Guam, the refuelling of all *Los Angeles*-class SSNs to extend their service lives, and the conversion of four *Ohio*-class SSBNs into conventional missile-launching and special operations forces-capable SSGNs, it is likely that the 55 submarines planned to be in service in 2015 will prove to be insufficient (Goure 2002).

American submariners also worry that their ability to perform their ASW missions will be eroded by the multiplying demands on the submarine force flowing from the new doctrinal priority given to operations in littoral waters. In addition to the SSNs' traditional missions of carrier group escort, intelligence, surveillance and reconnaissance and ASW, US Navy attack submarines are increasingly called upon to project force ashore in the form of the insertion of special operations forces or cruise missile strikes against remote land targets (Richardot 2002: 96). As Cote and Sapolsky (2001: 10) note:

> This 'multi-mission pull' increasingly makes ASW compete with strike warfare and theater air and missile defense for the same resources and training opportunities. The other mission areas are winning these battles and pulling the Navy's major platform communities away from ASW, particularly in the aviation and surface warfare branches.

Thus, in addition to the potential deficiencies of the US Navy's submarine-based ASW capabilities, its airborne ASW capabilities are also likely to be stretched by the increasing emphasis on the conversion of its previously dedicated ASW platforms to perform multi-role missions. For example, the US Navy's *Orion* P-3 maritime patrol aircraft, which have been hitherto optimised for maritime patrol and ASW missions, proved successful in

Afghanistan and Iraq as reconnaissance and communications relay aircraft. They are also being fitted with stand-off land attack missiles to enable them to undertake overland strike operations (Polmar 2004a: 88). At the same time, the P-3 force is being sharply reduced: at the beginning of 2004, the US Navy had some 225 P-3C *Orions*, but because of airframe fatigue and corrosion, around half of these aeroplanes were to be retired within a year (Polmar 2004a: 88). This means that with fewer aircraft available for ASW training missions, it will become increasingly difficult for ASW operators to exercise their skills, leading to reductions in their capabilities. The replacement of the P-3C fleet by the 108 new Boeing Multi-mission Maritime Aircraft (MMA), due to become operational in 2013 (Cavas 2004a: 6), is hardly likely to assuage the concerns of the ASW operators about the potential detrimental effects on their skills of no longer having a dedicated ASW platform.

Ship-based airborne ASW capabilities are also likely to suffer as a result of the replacement of the US Navy's SH-60B and SH-60F *Seahawk* helicopters with MH-60R *Seahawk* and MH-60S *Knighthawk* helicopters. The primary role of the SH-60B helicopters, which operated from surface combatants, and the SH-60F helicopters, which operated from aircraft carriers, was anti-submarine warfare. Their replacements will have multiple roles, including surface surveillance and attack, mine countermeasures, combat search and rescue, support for special forces operations and vertical replenishment (Polmar 2004b: 88). The multiplication of the roles of these helicopters means that, like the P-3, their availability for ASW training missions will be limited, leading to an erosion of ASW skills. The US Navy's ship-borne ASW capabilities have suffered from the reduction in the number of its Surveillance Towed Array Sensor System-equipped T-AGOS ships. Only five of the US Navy's 22 Cold War era T-AGOS ships continue to perform ASW missions (Goldstein and Murray 2004: 184).

The US Navy's efforts to improve its anti-submarine warfare capabilities to enable it better to meet the challenges of operations in littoral waters continue to be beset with problems. A May 2001 General Accounting Office (USGAO 2001b: 11) report found that:

> Although the Navy is making some progress in overcoming shortfalls identified in the 1997 Anti-Submarine Warfare Assessment, a lack of resources and priorities among competing programs is still prevalent. Funding reductions within the MK-54 Lightweight Torpedo program – the Navy's premier weapon against submarines in the littoral – will delay the fleetwide introduction and reduce the number of torpedoes the Navy can buy each year. Technical problems and cost growth have adversely affected the SH-60R helicopter conversion program that will work together with Navy ships to detect, track, localise and destroy enemy submarines. This

program's high cost has already forced the Navy to reduce the number of helicopters it intends to convert. The Navy has still not established priorities among individual antisubmarine warfare acquisition programs which would allow it to concentrate resources on the systems which would produce the highest payoff in added capability ... The Navy is pursuing several training initiatives to improve proficiency of crews. However, the shallow-water training ranges the Navy says it needs may not be available for many years, owing to funding limitations.

In recognition of the threat that conventional submarines pose to its forces operating in shallow and near-shore waters, the US Navy established a Task Force ASW in February 2003 to examine fleet shortcomings and recommend improvements to ASW capabilities. Acting on a recommendation of the Task Force, the US Navy has established a new Fleet Anti-Submarine Warfare Command. The avowed aim of the US Navy in taking these measures is to eliminate the dangers posed by conventional submarines and give US forces undersea maritime supremacy – just as US air power dominates the skies (Sherman 2003: 6).

8

THE UNIVERSAL AND THE PARTICULAR IN STRATEGIC LOGIC

We have tried so far to elucidate the reasons behind the prominent place of submarines in China's naval force structure, examining Beijing's strategic ambitions and the means available to achieve them and using a rational-actor type analysis and maritime strategic theory as a way of shedding light on the strategic choices of the Chinese leadership. But it would be a mistake to assume that these choices were simply the outcome of a process of reasoning by abstract rational actors based solely on a universal strategic logic.

The examination of certain features of Chinese strategic culture throws an alternative, complementary light on the reasoning which leads Chinese strategists to accentuate the role of submarines in the PRC's naval strategy. Chinese strategic culture is the product of China's unique historical experience, not only since the foundation of the People's Republic in 1949, but also during its 4,000 years of history preceding the formation of the most recent Chinese state. Many analysts would agree that there is an essential continuity in Chinese strategic culture and that this is an important factor in shaping Chinese strategic behaviour from ancient times into the present era. Andrew Scobell (2003: 16), for example, considers that 'there does not appear to have been any radical change in strategic culture certainly since 1949, and probably not for hundreds – perhaps thousands – of years'. This strategic culture forms the perceptual lens through which Chinese strategic policy-makers view the universal and timeless logical process of matching strategic ends and means.

According to Ken Booth and Russell Trood (1999: 8), the idea of strategic culture assumes that there exists within a state or nation or politically relevant group 'a distinctive and lasting set of beliefs, values and habits regarding the threat and use of force, which have their roots in such fundamental influences as geopolitical setting, history and political culture'. Booth and Trood maintain that 'these beliefs, values and habits constitute a *strategic* culture which persists over time, and exerts some influence on the formation and execution of strategy'. They (ibid.: 12) emphasise that 'strategic culture "shapes" but does not "determine" strategic behaviour'.

Strategic culture produces tendencies, it creates predispositions, but it does not determine policy. Strategic culture helps constitute attitudes and behaviour, but cannot on its own fully explain outcomes, since other variables, such as technology, play a part and may at any one point dominate.

In analysing a state's strategy, it is helpful to consider it as being composed of two interlocking parts, which they call 'statist military logic' and 'national strategic traditions' (ibid.: 13). The former component of national strategy is the dominant strategic logic of the interstate system as it has evolved over the past three and a half centuries. To the extent that all modern states participate in this system, the rational – albeit boundedly rational – instrumentalist strategic logic that inheres within the system is universal. Many theorists would go further and extend the military logic which Booth and Trood see as a product of the post-Westphalian state system to all political systems irrespective of historical time or geographic place. These theorists would agree with Gray (1999b: 9) that there is a 'unity of strategic experience across boundaries of time, place and technology' and that 'although tactical forms of war alter with political, economic, social and technological change, war and strategy retain their integrity as distinctive phenomena'. From this unity of experience arises a basic logic of strategy and statecraft which is also universal. To deny the universality of this strategic logic would be, in Michael Handel's words (2001: xvii), 'akin to asserting that Russia, China, Japan and the United States each follow distinct theories of physics or chemistry'. The fact that it is possible to discern a common logic in the prescriptions for the conduct of war by Sun Zi, Thucydides, Machiavelli, Clausewitz, Jomini and Corbett suggests that this military logic goes much further back in human history than the formation of the modern European state system.

Booth and Trood's (1999: 14) second component of national strategy, 'national strategic traditions', operates together with this universal military logic and, to a greater or lesser extent, alters it to produce variations on its basic themes. These variations are particular to each nation:

> *National traditions* consist of those peculiar epistemologies, outlooks and habits which create the 'common sense' which is particular to a distinctive identity group ... National traditions are shaped and perpetuated by 'who we think we are' in the context of the special historical experiences and geopolitical situations which we inherit. This is where strategic culture comes in. It constructs and helps us to account for those ideas which comprise a nation's common sense in different contexts pertaining to the threat and use of force.

Thus strategic culture offers a complementary dimension to explaining national patterns of thought and action, or why nations have 'preferred ways of coping with strategic problems and opportunities' (Gray 1986: 37). The analysis of strategic culture is, in other words, a useful tool to examine what J. C. Wylie (1967: 6) calls 'the patterns of thought that the military mind does use' – specifically in this case, the Chinese military mind – as a basis for some informed speculation on some patterns of thought that it might use.

Logic of consequences and logic of appropriateness

Booth and Trood's concept of strategy being composed of two complementary parts, one more general and objective, the other more particular and subjective, recalls the distinction between the rational actor and the organisational behaviour models of decision-making made by Graham Allison and Philip Zelikow in their classic analysis of the decision-making processes surrounding the Cuban missile crisis, *The Essence of Decision*. Allison and Zelikow noted that at the heart of the difference between the rational actor and the organisational behaviour models was the distinction between two different 'logics of action': a logic of 'consequences', and a logic of 'appropriateness'. To explain the difference between these two logics, they quote (1999: 146) James March and Herbert Simon, the organisational behaviour theorists who first drew attention to them:

> The first, an analytic rationality, is a logic of consequences. Actions are chosen by evaluating their probable consequences for the preferences of the actor. The logic of consequences is linked to conceptions of anticipations, analysis, and calculation. It operates principally through selective, heuristic search among alternatives, evaluating them for their satisfactoriness as they are found.
>
> The second logic of action, a matching of rules to situations, rests on a logic of appropriateness. Actions are chosen by recognizing a situation as being of a familiar, frequently encountered type, and matching the recognized situation to a set of rules ... The logic of appropriateness is linked to conceptions of experience, roles, intuition, and expert knowledge. It deals with calculation mainly as a means of retrieving experience preserved in the organization's files or their individual memories.

The 'logic of appropriateness' is the dominant logical mode of the second component of national strategy posited by Booth and Trood, that of 'those peculiar epistemologies, outlooks and habits which create the "common sense" which is particular to each national group' (ibid.: 14).

Thus the logic of consequences is the dominant operative logic in the process by which Chinese strategists have determined that a submarine-heavy naval force structure represents the best possible match between strategic ends and available means. This process involves the analysis of numerous variables including China's geographic and strategic environment, its economic circumstances and calculations of the balance of strength between Chinese forces and those of potential adversaries. The strategic choices which result from this process are guided, validated and reinforced by a complementary decision-making process in which the logic of appropriateness is dominant. This second process draws on elements in China's national experience of warfare that have produced ideas and patterns of behaviour, models of organisation and strategic problem-solving which make the submarine seem the most appropriate instrument to fulfil the strategic purposes of the Chinese state. In other words, within the conceptual framework of China's strategic culture, it makes 'common sense' for submarines to be accorded great importance in the structure of China's fleet.

This is not to say, however, that Chinese strategic culture is composed of elements that are purely Chinese, nor that the logic of appropriateness draws only on experience, knowledge or models of action that are purely indigenous. Chinese strategists and military planners belong to a transnational epistemological community of professional counterparts who share a common heritage of strategic thought and whose interests and experience lead them to develop their understanding and professional competence from a common pool of knowledge. Chinese strategists draw on wide and varied sources of theoretical and technical inspiration. Swanson records (1982: 199), for example, how in a college for senior naval officers established in Nanjing in 1955 students were reading a translated version of Mahan's *The Influence of Sea Power on History*. Corbett's *Some Principles of Maritime Strategy* was translated into Chinese in 1958 (Coutau-Bégarie 2000: 554). Mao Zedong's debt to Clausewitz has been well documented (e.g. Heuser 2002; Handel 2001). Shambaugh (2002: 3–8) has described how PLA strategists meticulously analyse each of the major campaigns in which United States military forces have been engaged since the end of the Cold War, and how these analyses have profoundly influenced the development of the PLA's basic and operational doctrines. PLA strategic planners are also avid readers of US armed forces' doctrinal and technical publications and journals and follow closely the debate on strategic issues among Western defence experts. As one group of Western PLA experts has observed,

> especially when it came to doctrine, the PLA quickly evolved into an eclectic 'learning organization' that adopted foreign theory, accepted what was useful, rejected what was not, and adapted to what they would refer to as 'the subjective realities of the situation'.
> (Ryan *et al.* 2003: 13)

9

INFLUENCE OF THE SOVIET
EXPERIENCE ON THE PRC'S
MARITIME STRATEGY

Of all the foreign influences on the PLA Navy, perhaps none is greater than that of the Soviet Union. As Shambaugh points out (2002: 108), the organisational structure of the PLA as a whole is still closely modelled on that of the Soviet military of the 1950s with a Central Military Commission, general departments, military regions and districts, and configuration of services.

The influence on the PLA Navy of Soviet and Russian technology and strategic concepts has, if anything, increased since the dissolution of the Soviet Union, particularly as a result of the embargo on the provision of Western military technology since the Tiananmen events in 1989. While the recent evolution of PLA doctrine as a whole has been shaped by Chinese strategists' scrutiny of American campaigns from the Gulf War through to the present war in Iraq, the lessons of these conflicts have less relevance for the PLA Navy. Control of the sea was an essential prerequisite for American and allied military forces to engage and prevail in every one of the post-Cold War conflicts, but in no case was this control seriously contested. Thus the Soviet experience, and Soviet strategies and doctrines forged during the Cold War, retain a currency for the PLA Navy that is not shared by the other services of the PLA for whom the lessons of the air and land battles of the wars against Iraq, Serbia and Afghanistan have greater relevance.

If anything, the lessons of the post-Cold War conflicts reinforce the relevance of Soviet naval strategy for today's PLA Navy. These recent conflicts have demonstrated the increasing reliance by US military forces on long-range, precision attack from sea-based platforms. During most of the Khrushchev era (1953–64) the primary strategic threat faced by the Soviet Union was from US aircraft carrier strike groups. Beginning in 1952, after a six-year post-war pause, the United States began a programme to construct a series of large, modern attack carriers. By 1958, the United States had seven new carriers under construction and three older carriers extensively modernised, and had begun to deploy long-range nuclear delivery aircraft such as the A-3D (Wolfe 1973: 252). With the introduction by the United States of the Polaris A-1 SLBM in 1961, the primary threat to the Soviet Union shifted from strike carriers to nuclear-powered ballistic missile submarines. The Soviet Navy's doctrine, operational strategies and force structure were largely shaped by the need to counter these threats. Since at least the Gulf War, and certainly since the deployment of the US Navy's two carrier battle groups to the Taiwan Strait in 1996, a central driving force behind the evolution of the PLA Navy's doctrine, strategy and force structure has been the need to counter US aircraft carriers. As the US military forces' operations in Iraq, Kosovo and Afghanistan have demonstrated, however, the US Navy relies increasingly on submarines to deliver long-range precision strikes against onshore targets. In Operation Iraqi Freedom in 2003, for example, the United States had 12 attack submarines in theatre, supplemented by two British submarines. Of the 800 Tomahawk missiles fired during this operation, one-third of them came from these boats (McKittrick 2003: 7). With the operational deployment in 2007 by the US Navy of its four converted *Ohio*-class SSBNs into conventional cruise missile armed SSGNs, each armed with up to 154 Tomahawk cruise missiles, the PLA Navy's threat perceptions could well undergo an evolution similar to that of the Soviet Navy in the early 1960s.

Formative influence of the Soviet model on the PLA Navy

Even before the fall of the Nationalist government on mainland China and the formal establishment of the PRC on 1 October 1949, a Communist naval academy had been established in Soviet-occupied Dalian in 1946 with a curriculum managed entirely by Soviet naval advisers. Within a year, the academy had graduated 300 midshipmen, many of whom received training aboard Soviet submarines operating from Port Arthur (Swanson 1982: 173). By 1951, the USSR had a large contingent of naval advisers in China to assist with the development of the PRC's navy: their number rose to around 2,000 by the time they were expelled in 1960 (Moore and Compton-Hall 1987: 199). The agreements signed by Mao and Stalin in Moscow in February 1950 provided for the supply of a broad range of maritime technical assistance, including instructors and technicians to help create a Chinese naval air force made up of aircraft provided by the USSR (Swanson 1982: 205). The submarine corps was established in April 1951, when a 275-person submarine study team, headed by Fu Jize, was sent to study in the Lushun Submarine Detachment under the Pacific Fleet of the Soviet Navy (Kondapalli 2001: 47). During the Korean War, the newly formed PLA Navy had to rely on Soviet ships and assistance (Cole 2003b: 158). At a ceremony in November 1951, Wang Hongkun, Deputy Commander of the Chinese Navy, stated: 'The Soviet Navy is an example for the Navy of the Chinese people, and is the direction of construction of the Chinese People's Navy. We should learn from the great Red Navy in order to speed up the building of a powerful people's Navy' (Swanson 1982: 193). During the 1950s, over 1,100 Chinese naval personnel were sent to the Soviet Union for training. Between 1956 and 1958, the PLA Navy undertook the translation into Chinese of 111 Soviet training manuals (Kondapalli 2001: 132).

The first commander of the PRC's new navy was Admiral Xiao Jingguang (1903–89), a contemporary of Mao Zedong. Xiao had attended a Russian-language school in Shanghai before being sent to study in Moscow from 1922 to 1924. He spent three further years in the Soviet Union after the split between the Kuomintang and the Communists in 1927 (Swanson 1982: 194). In 1955, Xiao began the process of modernising the naval establishment along Soviet lines with the adoption of the Regulations on the Service of Officers (Swanson 1982: 198). The debt that the Chinese Navy owed to Moscow was expressed by Xiao in a 1959 article:

> The development of the Chinese Navy is inseparable from the fraternal aid of the Soviet Union. At a time when our industry was still weak and our science and technology comparatively backward, the Soviet Government, the CPSU, and the entire Soviet people rendered us selfless aid.
>
> (Swanson 1982: 200)

Neither the Imperial Chinese Navy nor the Nationalist Navy had ever acquired submarines – although a small submarine was reportedly built under the Qing regime at Tianjin in October 1880 (Kondapalli 2001: xviii). It was the USSR that provided the first Chinese submarines. Among the first items of new naval hardware that the newly formed PLA Navy acquired were 50 P4 motor torpedo boats and one M-class short-range submarine, the latter delivered in 1953. In the spring of 1954, the USSR gave China three S-1 class submarines. By 1955, the PLA Navy was reported to have 4 M-type and 4 S-type Soviet submarines. One of the M5-class and two of the S1-class survived into the 1970s (Moore and Compton-Hall 1987: 195). At the same time, the Soviet Union gave China active support in building new warships, including modern *Whiskey*-class submarines at Shanghai's Jiangnan Shipyard, the first of which was completed in 1956 (Swanson 1982: 196). Between 1956 and 1964 the first five of 21 Soviet *Whiskey*-class boats were assembled in Chinese yards (Moore and Compton-Hall 1987: 195). China had also received Soviet plans for a new missile submarine, the *Golf*-class SSB, as well as for the Soviet *Romeo*-class submarine which was to replace the *Whiskey*-class when Soviet assistance ended in 1960 following the split between Beijing and Moscow. By 1963, the PRC had three yards in operation – at Canton, Shanghai and Wuzhang – building *Whiskey*- and

Romeo-class boats. The *Romeos* remained in production until the mid-1980s. With a total of 91 *Romeos* constructed, only the Soviet *Whiskey* programme produced more boats of the same class in the years after the Second World War (Moore and Compton-Hall 1987: 195).

Despite the next three decades of antagonism between Beijing and Moscow, the PLA Navy remained indelibly marked by its early dependence on Soviet training, technology and equipment. Following the split, China continued to pursue a Soviet-style naval construction programme. For example, nine more *Whiskey*-class submarines were launched between 1960 and 1964, and by late 1964, a follow-on prototype, a replica of the Soviet *Romeo*-class, was in production (Swanson 1982: 226). Although Chinese-built warships began to predominate in the PLA Navy for the first time in the 1970s, they were still heavily dependent on Soviet designs (Cole 2001: 95). Many of the officers who rose to the higher command levels of the PLA Navy, like Liu Huaqing, had received their advanced training in the Soviet Union.

With the restoration of good relations between Beijing and Moscow following the dissolution of the Soviet Union, Russia is once again China's most important supplier of advanced naval equipment, technology and training. Following the re-establishment of military-to-military relations and the initiation – at the time of Li Peng's visit to Moscow in 1990 – of negotiations over arms sales, during the following decade China is estimated to have bought weapons and equipment from Russia worth some US$6.75 billion (Shambaugh 2002: 218). Chinese naval officers have been training in Russian naval academies since 1994, and the Chinese and Russian navies have been conducting joint manoeuvres since 1999 (Kane 2002: 110). Russian engineers are known to be assisting in the design and construction of China's new Type-093 nuclear-powered SSN.

The fact that the PLA Navy was modelled on the Soviet Navy and that those who have been responsible for setting its strategic directions were so profoundly influenced by the Soviet example, was without doubt a significant element in the logic of appropriation infusing the strategic planning of the PLA Navy's force composition and structure. The decision to adopt a structure in which tactical submarines play a major role is likely to have been due in no small part to the influence of Soviet thinking and experience and the Soviet model of naval force composition.

The navy that the leadership of the PRC inherited from the defeated Nationalist government was characterised by inadequate, obsolete and defective equipment which had been neglected during years of civil warfare. The Nationalist Navy had been notoriously corrupt and more interested in piracy and smuggling than in fighting for the Nationalist cause. Training and discipline had also been neglected and morale had been undermined by the preceding years of political turmoil and the inglorious combat record of naval forces. These forces had in any case played only a minor role in the fighting which had mostly taken place on land. Moreover, the political allegiance of the few officers who had sided with the revolutionary forces was suspect. The deficiencies in the new Communist Republic's naval power became starkly apparent in October 1949 when the PLA attempted an amphibious assault on Quemoy (Jinmen) Island and were driven back by an estimated 15,000 Nationalist troops (Swanson 1982: 185).

The challenge in reconstructing and modernising China's naval forces which the leadership of the newly established PRC had to confront in the early 1950s must have seemed very similar to the challenge faced by the Bolshevik leadership in the early 1920s. Gorshkov (1979: 133) describes the situation facing Lenin in 1921 in terms which could equally well have described the situation faced by Mao in 1949:

> By that time the Soviet country had no fleet in the Far East and the North, nor did it really exist in the Black Sea since nearly all the remaining ships had been taken off by the White Guards to foreign ports. Only the Baltic fleet and a few naval flotillas existed. Many ships were badly in need of repair. Over half the command staff were officers of the old Czarist fleet. Ratings and junior commanders were in need of replacement. Soviet command forces needed theoretical and practical training.

Even if the Korean War had not intervened to ensure that Moscow became the only source of advanced technical and material assistance available to China, the similarity of experience between the two countries would have ensured that Chinese policy-makers regarded the Soviet model as the most appropriate to their particular circumstances. In the case both of the Soviet Union in the 1920s and China in the 1950s, the dire economic and strategic circumstances facing the new revolutionary governments precluded any major investment in new capabilities for their navies, even if their national industrial infrastructures had been capable of producing sophisticated modern platforms and weapons. Neither the Chinese political and military leadership nor the Bolshevik leaders were in any case convinced of the need for large or well-equipped naval forces. In both cases, the military hierarchy was dominated by army officers with little understanding of the importance of sea power, and the military tradition of both Russia and China was that of land warfare. Mao Zedong's attempt to derive naval doctrine and strategy from his experience of partisan warfare echoed those of Frunze, Tukhachevskii, Vorishilov and other Bolshevik military leaders who sought to shape Soviet maritime strategy to reflect their Russian Civil War experiences (Herrick 1968: 19). And in both cases political and military leaders had an acute sense of being encircled by hostile alliances of states bent on effecting a 'regime change' in the new Communist state. It is therefore not surprising that the strategic doctrines and force structures of the PLA Navy in the 1950s and 1960s should so closely resemble those of the Soviet Navy in the 1920s and early 1930s.

The Soviet naval forces of the 1920s and early 1930s were constructed in conformity with a new strategic doctrine which took account of these difficult strategic and economic realities. This doctrine, known as that of the 'New School' (sometimes known as the 'Young School'),[1] rejected the idea of a high seas fleet made up of major surface combatants and proposed that the navy should instead concentrate its efforts on inshore defence using light surface craft, submarines, mines and shore-based aircraft. The basic tenet of the Soviet 'New School' was, according to Robert Herrick (1968: 21–2), 'simply an assertion that the submarine had replaced the battleship as the main striking unit of the fleet … Submarines, aided by aircraft and light surface ships and craft, were held to be the major weapons of contemporary naval warfare.' Thus it was that the first Soviet naval construction programme tabled in 1926 contained a high proportion of submarines (Moore and Compton-Hall 1987: 67).

As the Soviet economy and industrial base recovered from the devastation of the First World War, the Revolution and the Civil War, Moscow was able to initiate a large-scale submarine production programme in 1933 which resulted, by 1939, in the Soviet Union having the world's largest submarine force (Ranft and Till 1983: 86). The Soviet Navy, and its Russian successor, maintained this quantitative lead for the remainder of the twentieth century. By the early 1980s, the Soviet Navy's submarine tonnage accounted for nearly one half of its total naval tonnage. While the US Navy had built some 150 submarines in the four decades since the end of the Second World War, the Soviet Navy had built around 600 (Ranft and Till 1983: 108). In the 1950s, at the time when Soviet influence on the newly formed PLA Navy was at its height, the Soviet Union was engaged in a massive submarine building programme adopted in 1948 with the goal of adding 1,200 submarines to the Soviet fleet between 1950 and 1965. The plan called for the construction of 200 long-range *Zulu-* and *Foxtrot*-class boats, 900 medium-range *Whiskey-* and *Romeo*-class boats and 100 coastal *Quebec*-class submarines (Moore and Compton-Hall 1987: 66). This massive commitment on the part of the PLA Navy's mentors to a naval force structure heavily weighted towards submarines could not have failed to have shaped the thinking of Chinese naval planners.

Thus, during the first decades of its existence the maritime strategy adopted by PLA Navy was closely modelled on that of the Soviet 'New School'. Like that of its Soviet antecedent, the central element of this strategy was one of coastal defence by small surface units, shore-based planes and submarines (Cole 2001: 161). In sum, it was a strategy which Chinese strategic planners had arrived at by means of a logic of appropriateness, identifying in the Soviet model a strategy which had been fashioned for circumstances very similar to those which the leadership of the newly founded PRC faced in the 1950s (Cole 2001: 286). The doctrine formulated by Xiao Jingguang in 1950 was a copy of the Soviet 'minor war' theory which required, in Xiao's words: 'a light type navy, capable of coastal defence. Its key mission is to accompany the ground forces in war actions. The basic characteristic of this navy is fast deployment, based on its lightness' (You 1999: 164).

This doctrine determined the PLA Navy's force structure which was based on torpedo boats, land-based naval aircraft and submarines. As in the Soviet Navy, the submarine fleet was given a particular priority. The PLA Navy continued to adhere to the model of the Soviet 'New School' strategy even after the rupture between Beijing and Moscow in 1959. During the 1960s and 1970s the strategic mission of the Chinese Navy was restricted to providing support for ground forces, controlling criminal activities such as piracy and narcotics trafficking, safety and rescue, and ensuring security of navigation. The only additional role came with the development of a strategic deterrence capability through the construction of the *Xia* SSBN and its JL-1 missile.

In the meantime, after the Cuban missile crisis had demonstrated the extent to which Moscow's ability to achieve its strategic objectives was handicapped by its lack of long-range sea power, Soviet naval strategy underwent a thorough revision under the direction of Admiral Gorshkov. The new challenge for the Soviet Navy was to construct a blue-water fleet to rival that of the United States. The wartime missions of this fleet would be the defence of areas offshore from the Soviet Union, opposition to enemy strategic strike systems, command of the sea in the areas of operation of the Soviet SSBNs, strategic nuclear attack, breaking enemy sea lines of communication and protecting those of the Soviet Union. In peacetime, the Soviet Navy added the roles of naval diplomacy, crisis management, support for Soviet economic interests and local sea control.

For China, the shift to a more outward-looking strategy had to await the passing of the upheavals of the Cultural Revolution, from which only the priority nuclear and ballistic missile programmes had remained relatively unscathed, and Deng Xiaoping's emergence from the leadership struggles following Mao's death in 1976 to head the Chinese political hierarchy. Deng's appointment of Liu Huaqing to command the PLA Navy was an indication of his determination to enhance China's sea power, since Liu's substantive rank (general/admiral) was more senior than that normally held by PLA Navy commanders (lieutenant-general/vice admiral) (Cole 2001: 72). But Liu's formative exposure to Soviet naval experience and teaching, and the attraction that Gorshkov's ideas had for him, ensured that the Soviet model would continue to exert a strong influence on the evolution of Chinese naval doctrine and force structure despite the 30-year period of hostility between the two countries.

Admiral Sergei Gorshkov

The central message of Gorshkov's book *The Sea Power of the State* was that no country could aspire to be a great power unless it was strong at sea. The failure to understand this essential truth by all of Peter the Great's successors had, according to Gorshkov, resulted in the repeated humiliation of Russia by foreign powers. This was a theme which no doubt struck China's nationalistic Communist leadership as equally relevant to China's experience during the preceding century, in which the country had been repeatedly humiliated by the European and Japanese imperialist powers. Liu would no doubt also identify with the challenge that Gorshkov had so successfully met in convincing the army-dominated, continentalist-minded Soviet military and political hierarchy of the important contribution that an effective blue-water fleet could make to the attainment of the Soviet Union's strategic goals. When Liu took up the task of reforming the PLA Navy, he faced a similar challenge in arguing for the resources he would need to construct the blue-water navy that he set as the ultimate goal of his efforts. In 1982, when Liu was given this task by Deng, the outlook of the Communist military and political hierarchy was not very different from that attributed by Swanson (1982: 183) to the leadership in the early 1950s:

> at that time, the Chinese Communist leadership was still made up of people who sprang from the peasantry and had little formal education. They were trained guerrilla fighters, experts in riverine warfare, with no knowledge of navies and maritime matters.

Writing in 1982, just before Liu was appointed as Commander-in-Chief of the PLA Navy, Swanson noted that 'the Communist leadership is still divided on the question of modernization and agreement on future navy building is lacking. Because of this, the navy finds itself at a critical crossroads, cautiously examining its alternatives' (Swanson 1982: 285). Under Gorshkov's guidance, by 1982 'the Soviet Union had developed from a minor naval power with largely coastal capabilities [to] become a major sea power able to challenge the supremacy of the US Navy throughout the world' (Ranft and Till 1983: 206). At this juncture in the evolution of the PLA Navy, Gorshkov must have been a powerful inspiration to Liu Huaqing: it is thus not surprising that he should have recommended to his colleagues that they should read Gorshkov's book.

Gorshkov was a firm believer in the value of submarines, despite his being best known as a strong advocate of a 'balanced fleet' integrating land- and sea-based aircraft, surface combatants and strategic and tactical submarines. Like Castex and Doenitz, he considered that in the First World War Germany had failed to make the best use of its submarine fleet which, had it been used more effectively, could have been a decisive weapon

(Gorshkov 1979: 100). In the Second World War, he considered that Germany had underestimated the contribution that submarines could make towards the achievement of its strategic objectives, and that 'aircraft and submarines were assigned by the allies, like the Germans, a secondary role for which they had to pay dear in the course of the war' (ibid.: 108). For Gorshkov (ibid.: 123), the central lessons for naval fleet composition and structure of the Second World War were the importance of constructing balanced fleets, the obsolescence of battle fleets and the passing of the primacy among surface ships to aircraft carriers, and that 'submarines became the most important means of battle at sea'. Submarines, according to Gorshkov (ibid.: 190), 'sank more merchant and fighting ships than did surface ships and aircraft taken together'.

According to Gorshkov (ibid.), one of the primary reasons for the Soviet Union's decision after the Second World War to make its submarines 'the main combat power' of its fleet were the 'military-economic considerations' which encouraged the Soviet fleet to switch over to building primarily submarine forces. These 'military-economic considerations' are undoubtedly an equally important factor in China's decision to develop a powerful submarine arm for its fleet. As Gorshkov (ibid.) explains it:

> In determining the lines of development of the navy in the nuclear age, one could not fail to take into account, for example, the fact that the imperialist states opposing us possess an enormous surface fleet and a powerful ship-building industry. Even for us to match in numbers the main classes of surface ships would have taken us many years of estimating the relative potentials, involving the expenditure of enormous material and monetary resources ... The priority given to the development of submarine forces made it possible in a very short time to increase sharply the strike possibilities of our fleet, to form a considerable counter-balance to the main forces of the fleet of the enemy in the oceanic theatres, and, at a cost of fewer resources and less time, to multiply the growth of sea power of our country, thereby depriving an enemy of the advantages which could accrue to him in the event of war against the Soviet Union and the countries of the socialist community.

For a country such as China, weaker than its potential adversary not only in terms of naval power but also in terms of financial power, this is also a powerful argument in favour of submarines. In the mid-1990s, China acquired four *Kilo*-class submarines from Russia for a cost of around US$1 billion (Shambaugh 2002: 273); the first *Sovremmeny* destroyer was bought in 1996 for a reported price of US$840 million and the second in 1997 for US$1 billion (ibid.: 267). For the Chinese leadership, quite apart from any asymmetrical operational advantages that submarines can offer an inferior navy, the fact that for the price of one *Sovremmeny* destroyer China can buy four *Kilo*-class submarines is not a trivial reason for favouring submarines over surface combatants.

Gorshkov (1979: 191) described the Soviet Union's post-war development of its submarine forces as having two stages: in the first stage, lasting about ten years, 'improved submarines of new designs were created with diesel-powered engines. In the second stage ... mostly submarines with atomic-powered systems are being built.' China is following a similar path, although the transition from the first stage of conventionally powered to the second stage of nuclear-powered submarine construction has been considerably delayed by the withdrawal of Soviet technical assistance at the end of the 1950s, which forced Chinese scientists and technicians on to a difficult, time-consuming, expensive, trial-and-error path of indigenous construction of its nuclear-powered submarines. China's first nuclear-powered submarine, the *Han*-class attack submarine (SSN), was begun in 1958 on the basis of a Soviet design but did not go to sea until 1974 (Cole 2001: 98). It did not reach full operational status until the mid-1980s (Shambaugh 2002: 272). Its single *Xia*-class SSBN joined the fleet in 1988 after 16 years of development (ibid.: 171). The development of its replacement, the Type-94 class SSBN, began in the late 1980s, but has reportedly encountered problems which are likely to delay its deployment into the next decade (ibid.: 272).

The Chinese have also followed the Soviet example, rather than that of the United States, in not abandoning diesel-electric submarines in favour of nuclear-powered boats, no doubt because the PLA Navy, like the Soviet/Russian Navy, values the relatively low cost of the latter compared with the former, as well as their ability to operate quietly and effectively in shallow waters (Ranft and Till 1983: 113).

The Soviet model of defence against maritime attack

Gorshkov (1979: 66) maintained that because of the political and historical conditions of its genesis, the development of the fleet of Russia – the largest continental state in the world – was quite distinctive. Chinese naval strategists would see a particular appropriateness in the logic which resulted in the strong emphasis of Soviet naval strategy on defence against maritime attack. The circumstances which produced this traditional emphasis in Russian and Soviet naval strategy correlate closely with those which generate Chinese anxieties about the potential for a maritime attack by the United States and its allies. Chinese strategists' analysis of the Gulf War and subsequent campaigns in which the United States used sudden and overwhelming force against vital points of the enemy's defence assets and infrastructure would lead them to assume that in the event of a war over Taiwan, US forces would conduct a similar campaign of crippling strikes against critical

Chinese targets. The key objective of the Chinese defence strategy would therefore be to prevent US forces from carrying out such strikes, or at least to minimise their effect. Thus, the traditional emphasis in Soviet naval strategy of defence against maritime attack, with its interest in minefields, submarines and coastal artillery, has an obvious contemporary relevance for Chinese naval strategy.

The most important single lesson, according to Herrick (1968: 55), 'that could be learned from Soviet naval participation in World War II is that of the superiority of the tactical offensive over the defensive, even on the part of fleets forced to take a strategically defensive posture'. Thus the World War II experience only served to reinforce the central idea in the Soviet 'New School' strategy of the dispersion of Soviet naval power between such large numbers of submarines, aircraft and torpedo boats that an incoming invasion force would be subject to an attack of increasing diversity and intensity the nearer it came to Soviet territory (Ranft and Till 1983: 154). Like Liu Huaqing's concept of a progressive, triple-phased expansion of China's naval capabilities to cover maritime areas stretching further and further from Chinese shores, Soviet maritime defence concepts were 'based on three concentric areas extending out from their coasts, with the depth of each area based on the military capabilities of operationally available weapons systems' (Herrick 1968: 137). Thus, following the Second World War in which the United States and its maritime coalition partners had demonstrated so convincingly their ability to carry out successful amphibious assaults, Soviet naval strategists developed a plan to counter maritime attacks which, as Ranft and Till (1983: 154) describe it, involved the preparation of :

> a defensive perimeter up to about 500 miles from the Soviet coast within which invading forces would come under increasingly severe attack as they headed for Russia. Beyond and at the outer limits of this perimeter, defending forces would be restricted to land-based bombers, long-range submarines (*Zulus*) and a few heavy ships (*Sverdlovs*, *Chapaevs* and various destroyers) whose exact function was not very clear. Closer in, the stress would be on huge numbers of medium-range and coastal submarines (*Whiskeys* and *Quebecs* respectively), more and more aircraft, a swarm of minor combatants, especially torpedo boats, dense minefields and last but not least, powerful batteries of coastal artillery. The idea was that each of the four fleets would be self-supporting and this required the vast number of submarines (about 1,200) which the Soviet government claimed to be building in 1948.

This concept acquired added impetus with the development of Western carrier-based aviation with its progressively increasing ability to conduct strikes against targets further and further inland from launch locations further and further out to sea. The Soviet response to this threat was to develop the means to destroy the carriers before they could launch their attacks.

The Soviet Navy developed various tactics – based on those of the 'New School' – of massed and orchestrated attack by aircraft, submarines and surface ships designed to overwhelm the defences of the carrier task groups by sheer weight and diversity of numbers. In its Okean-70 exercise, for example, the Soviet Navy simulated a concerted attack on a carrier task group passing through the Greenland–Iceland–United Kingdom gap by 10 missile-armed surface ships, 30 submarines and 400 aircraft sorties, operating in waves (Ranft and Till 1983: 156). As the strike range of Western carriers began to outstrip the range of Soviet land-based aircraft, the Soviet defence perimeter was pushed out to some 1,500 miles, and the anti-carrier role of submarines became increasingly important in Soviet strategy (Ranft and Till 1983: 156). Thus it was that the principal target of the Soviet attack submarine fleet became the US Navy's aircraft carriers and, from the 1960s, American ballistic missile submarines (Van der Vat 1995: 324), just as, in the brief period in 1916 before Germany redirected its U-boat fleet towards waging a *guerre de course* in the Atlantic Ocean, Admiral Scheer had sought to use his submarines in a war of attrition in the North Sea against the capital ships of the opposing Grand Fleet.

With the adaptation of its operational doctrine to take account of the requirement to fight 'naval warfare under high tech conditions' made evident by the 1991 Gulf War, the PLA Navy introduced a number of new subjects into its training curriculum, including ' two vessels taking turns in attacking a large isolated naval target, an individual submarine launching sustained attack on multiple targets, submarine-to-aircraft confrontation, and long-range torpedo attacks' (Kondapalli 2001: 135) – subjects which are well suited to preparing personnel for the kind of anti-carrier tactics developed by the Soviet Navy in the 1960s and 1970s. Like the Soviet Navy before it, the PLA Navy is also preparing to engage hostile forces at increasingly greater distances from the shore. As far back as 1976, according to Kondapalli (ibid.: 146), the CMC approved for the first time an oceanic training voyage for the personnel of Submarine No. 252, the first to go beyond the first island chain into the open Pacific Ocean. Several naval bases have also carried out new tactical training programmes in seaborne refuelling of submarines and ships (ibid.: 147). The PLA Navy has also been conducting its training exercises at increasingly greater distances from the Chinese coast, and for increasingly longer periods (ibid.: 158).

Norman Friedman (2000: 129) argues that the Soviets pioneered the new information-centric style of naval warfare which is now identified with space systems, and that the genesis of this style of warfare, so naturally suited to coastal defence strategies, was in the system devised by the Germans to defend the Flanders coast occupied by them during the First World War.

The system used radio to link shore artillery, offshore craft (including submarines) and aircraft to a shore-based central command. Offshore scouts were positioned to detect and relay information about the approaching enemy fleet to the commander who would then cue coastal artillery, fast torpedo craft and other combat units into position to attack. Friedman suggests that the Soviet military, who had close relations with their German counterparts in the 1920s, adapted this system to their own situation. He (ibid.: 336n) also suggests that the system may have inspired Doenitz' operational approach to the U-boat war two decades later.

Friedman (ibid.: 130) observes that the system, which gave the central shore-based commander full control of all the craft assigned to him, was at odds with traditional naval practice which emphasised the initiative of each ship's commander. But it was attractive to the Soviets for political reasons because of the suspect loyalties of the ex-Tsarist officers on whom the new Soviet state had to rely. Central command made it more practical to exert political control by means of the commissars, who had to countersign commanders' orders. The system was also tactically advantageous in that it permitted coordinated attacks which increased the hitting rate of multiple small craft while reducing their vulnerability.

The system would clearly have a natural appeal to the PLA Navy – from both a political and a tactical perspective. The PLA Navy is organised on the model of the Soviet Navy with a dual system of political and military command up to the highest level of the service: the PLA Navy commander and the political commissar are usually of similar rank (Cole 2001: 73). In the early years of the PLA Navy at least, the commissar had to approve the commanding officer's orders, although with the increasingly complex and technical systems which characterise the modern Chinese Navy, the role of the commissars in military decision-making has been reduced and the focus of their responsibilities has shifted to essentially non-operational areas (ibid.: 135).

If the Soviet model of a defensive naval strategy (at least in its 1950s and 1960s variant when Moscow regarded carrier-based strikes as the principal threat) is reflected in China's planning for a response to a possible intervention by US forces during a Taiwan Strait crisis, this could involve the deployment of its more capable *Kilo-*, *Song*-class diesel submarines, and possibly its *Han*-class nuclear attack boats, to the outer limits of a defensive perimeter, and its less capable *Ming* and *Romeo* boats deployed closer in, operating in conjunction with surface combatants, aircraft and coastal missile batteries. This kind of deployment reflects the lessons that Castex drew from his analysis of the operational experience of Admiral Scheer's U-boats in 1916. Castex (1997: vol. I, 314) concluded:

> [The submarine] opens new horizons for strategy, by enabling a richness of operational possibilities infinitely greater than those

which would result from the simple application of the principle of concentration of all one's forces in the same place. But, to fulfil its promise, it should also be manoeuvred skilfully. It should not be used in a single linear barrage of submarines, which would have little chance of achieving something useful, but, on the contrary, one should, as Admiral Scheer wanted to, create an in-depth formation allowing for continuity of effort and the possibility of guarding against the unforeseen.

Just as the evolution of Soviet naval strategy was accompanied by a doctrinal tension between the advocates of the 'New School' doctrine of a defensive posture based on inshore and land-based defences and those, like Gorshkov, who advocated the development of a more distant, blue-water defensive capability, there is evidence today of similar doctrinal debates within the Chinese defence community. Several Western analysts have described this doctrinal debate which divides Chinese strategic planners into three camps: the 'Local War' advocates, the 'RMA' advocates and the 'People's War' advocates (e.g. Dreyer 2000; Hawkins 2000; Pillsbury 2001). The first two schools both advocate reforming the PLA and providing it with the appropriate capabilities to enable it to meet the challenge of the post-Cold War strategic environment and effectively counter an attack by US armed forces, which they consider to be the principal threat to China's national security. The former favours a gradual process of modernisation of China's armed forces to develop power-projection capabilities, building on its current capabilities, to enable the PLA to fight a limited local war under high-tech conditions; the latter advocates the acquisition of new advanced technologies, doctrine and organisational structures which would enable China to leap-frog beyond the capabilities of the United States. The 'RMA' school appears to have achieved some success in advocating its views with Jiang Zemin's reported demand in August 1999 for the accelerated development of the so-called 'Assassin's Mace' weapons (Pillsbury 2001: 4). Both of these reformist schools are opposed by the conservative 'People's War' advocates who cling to the Maoist doctrine of relying on China's numerically strong, albeit technologically backward forces to repel a large ground invasion of China's national territory.

This debate, as reported by Western analysts, is cast in terms of spending and equipment and no doubt reflects the rivalry among the PLA ground, air, missile and naval forces for a greater share of the limited national defence budget. These analysts do not ascribe particular viewpoints in this debate to particular services, although it would be natural to assume that the conservative, 'People's War' school would find the largest number of adherents among the PLA ground forces, rather than among those services which are more technology- than manpower-intensive. It would not be surprising if, within PLA Navy circles, this broader debate about equipment and resources acquired the form of the classic debate among naval strategists of continental powers pitting the advocates of a close-in, reactive defensive strategy designed to exploit the inferior continental navy's comparative advantage in coastal waters, against those who advocate a more proactive offensive operational strategy of seeking out and attacking the superior navy's forces before they are able to launch attacks against the shore. In the Soviet Union the primary missions of naval forces were conceptualised, in Gorshkov's (1979: 213) words, as 'fleet against shore' and 'fleet against fleet'. Within the framework of this conception of the role of naval forces, the debate polarised into those who, like Marshal Grechko, saw the primary task of navies as defending against the enemy's 'fleet-against-shore' strategies (Ranft and Till 1983: 70) and those who, like Gorshkov, emphasised the value of adopting more offensive and pre-emptive 'fleet-against-fleet' operational strategies.

Given the strong influence – both historically and currently – of Soviet and Russian naval thought on China's naval strategists, it would not be surprising if within the PLA Navy Chinese naval strategists were divided between the advocates of a defensive operational strategy to counter the enemy's 'fleet-against-shore' actions and the supporters of a more offensive and pre-emptive strategy of a 'fleet-against-fleet' approach.

10

CHINESE STRATEGIC CULTURE – INDIGENOUS ELEMENTS

The classical tradition of Chinese military thought

For all that the PLA Navy's dependence on Soviet training, equipment, doctrine and experience was a critical factor in shaping its early development, instinctive Chinese wariness of the dangers of relying too heavily on foreign sources of technology and ideas ensured that the Soviet model was not adopted without a critical appraisal of its suitability for local Chinese conditions. Mao himself worried about historic Russian efforts to dominate China and inveighed against the blind acceptance of all things Soviet to the detriment of the nation's sovereignty (Lewis and Xue 1994: 3). Indeed, as Kondapalli (2001: 173) notes, by the late 1950s reaction to the wholesale adoption of Soviet techniques and models of development had emerged in the form of criticism of 'doctrinairism and formalism' in naval work. Peng Dehuai, who sought to copy the Soviet model of naval development 'uncritically', was forced to admit his error: 'While learning from the experience of other countries in creating a modernized army, unsuitable training and working methods were adopted without giving sufficient consideration to actual conditions' (ibid.: 174). Lewis and Xue (1994: 220) argue that the Chinese High Command adopted a doctrine of coastal defence more because of its inherent suitability for the PRC's geopolitical and economic circumstances at the time rather than because it came recommended by Moscow: 'that coastal defense echoed early Soviet naval thinking was a coincidence, an extra bonus, not a conscious choice'. While the senior military leaders who laid the foundations of the PLA Navy acknowledged the relevance of the Soviet experience, Lewis and Xue (ibid.) maintain that 'they consciously sought a principle grounded in China's actual circumstances'.

The most obvious source of indigenous inspiration for doctrinal development was China's 4,000-year tradition of military thought. In this respect, few would deny the validity of Chen-Ya Tien's (1992: 13) observation:

> As one of the oldest civilized countries, with more than four thousand years of history, China has one of the richest military legacies

in the world. The development of modern military technology, the exposure to foreign military theories, and the repeated defeats in wars against the Western powers, have broken the monopoly of the ancient military theories but they are still highly respected and continually influence the thinking of Chinese military leaders.

As Ken Booth (1979: 73) has written, 'strategic theories have their roots in philosophies of war which are invariably ethnocentric. National strategies are the immediate descendents of philosophies of war.' One could add that the most concrete manifestations of these national strategies are the decisions on force composition and structure through which they are put into effect.

Thus, in addition to foreign sources of ideas on strategic theory, contemporary Chinese strategists consciously seek to ground their approaches to strategic analysis and planning in their own traditions. Ever since the first academy was established in 1072 under the Song dynasty to train candidates for military examinations, the Seven Military Classics, including Sun Zi's *Art of War*, have been required reading for Chinese military officers (Johnston 1995: 4). The PRC has maintained this tradition: Marshal Liu Bocheng, Head of the Chinese Military Academy in the 1950s, set Sun Zi's *Art of War* as a textbook for future officers (Niquet 1999: 89). As Valérie Niquet (1997: 17) remarks, contemporary Chinese strategists, motivated in part by Confucian respect for the classics and in part by a nationalist impulse to seek indigenous sources of inspiration, search the works of Sun Zi, Sun Bin and Wu Zi for a strategic approach to present-day challenges which is rooted in Chinese history and tradition. In the presentation of her French translation of Sun Zi's *The Art of War*, Niquet (1999: 97) notes that

> conferences are regularly organised by the Academy of Military Science on Sun Zi's *The Art of War* and institutes for the study of Sun Zi have been established, notably within the Academy of Military Science and the National Defence University.

The military classics of ancient China, which date back to the end of the fifth or beginning of the fourth century BC, are thus an integral and active part of the intellectual armoury of modern PLA strategists.

Classical Chinese military thought therefore continues to exert a powerful influence on the approaches to strategic problem-solving of modern Chinese strategists. Mao Zedong's military thought owes a considerable debt to the ancient military classics, even though during the Cultural Revolution he denied having been influenced by Sun Zi. In his 1936 (1963: 85) essay on 'Problems of Strategy in China's Revolutionary War', he urged the careful study of the lessons of past war because 'all military laws and military theories which are in the nature of principles, are the experience of past wars summed up by people in former days or our own times'. Mao also frequently referred to the classical Chinese military writings to bolster the arguments in his own military writings. Niquet (1997: 47) notes that, in 1942, a work was published in the Communist base in Yan'an which brought together the classical writings of Sun Zi, Wu Zi, Sima Fa, Weiliao Zi and extracts of Qi Jiguang. This book was designed as a training manual for the Party's military cadres. Mao's own military writings, permeated deeply as they were by the classical Chinese strategic thought, served as an indirect means of transmitting this classical heritage to the current era and giving it contemporary relevance.

There is much in the military classics of ancient China which, although the products of an overwhelming tradition of land warfare, and addressed to commanders of ground forces, is nevertheless germane to the development of a strategy and doctrine for undersea warfare. Particularly relevant to submarine warfare are the extensive consideration given by the classical authors to situations in which a weak force confronts a stronger opponent, and their counselling of the use of surprise, ruses and stratagems and indirect approaches rather than direct frontal, force-on-force attack. Concealment, deception and surprise – the qualities which are implicit in the modern meaning of stealth – are, after all, what give the submarine its essential comparative advantage over surface combatants.

Strategies for the weak against the strong

Classical Chinese strategic thought gives considerable attention to the strategies and tactics relevant to conflict situations in which opponents are asymmetrically matched. Indeed, as Paul Godwin (2003a: 23) has noted, down to the present day,

> much of the continuity found in China's military doctrine, strategy and concepts of operations, especially adherence to the principles developed by Mao Zedong in the 1920s and 1930s, is to be found in the enduring requirement to defeat superior opponents.

A prominent theme in Sun Zi's *Art of War* is the turning of 'disastrous circumstances into advantageous situations' (Chow-Hou 2003: 172). *The Art of War* bluntly affirms that 'the strength of an army does not depend on the superiority of numbers' (ibid.: 266). In fact, Sun Zi provides advice on the principles for the deployment of forces across a range of circumstances from overwhelming superiority over opposing forces through to crushing inferiority:

> When outnumbering the enemy ten to one, surround him.
> When outnumbering the enemy five to one, attack him.
> When outnumbering the enemy two to one, divide him.

> When comparable in numbers to those of the enemy, it is possible
> to engage him.
> When lesser in numbers than the enemy, be capable of escaping.
> When greatly inferior in numbers to those of the enemy, be capable
> of avoiding him.
>
> (ibid.: 65)

In his annotations to the translation of this passage, Chow-Hou Wee notes that avoiding the enemy can also mean concealing oneself from the enemy by taking steps to avoid detection. This nuance has obvious relevance to the advantages the stealthy characteristics of the submarine can offer to inferior naval forces. Chow-Hou also points out that equality or inferiority in numbers does not necessarily prohibit offensive action. It is still possible to attack the enemy provided that steps are taken in advance to distract or divide his forces, and to ensure that a smaller force has the means to escape after launching a surprise attack. Again, the approach implicit in Sun Zi's advice has obvious relevance to submarine warfare.

Sun Zi quite pragmatically advises the weaker force to 'defend when forces and resources are inadequate' (ibid.: 90), and then goes on to say that 'the adept in defence is able to conceal his forces in the most secretive ways and places in the earth' (ibid.: 92). In a later chapter, Sun Zi advises 'when in concealment (of forces and positions), be as inscrutable as the darkness of night' (ibid.: 186). Again, in the context of naval warfare, this advice evokes the historical preference of weaker naval powers for submarine forces because of the ability of these vessels to conceal themselves below the surface of the sea.

Similarly, unconventional actions which are secretive, deceptive, unpredictable and beyond imagination are also a recommended element of the repertoire of the offence: 'the adept person at offence is able to deploy his troops in ways that are beyond the imagination of anyone' (ibid.: 93). By acting in this way, according to Sun Zi, 'he is not only able to ensure the greatest security for himself, but is able to secure the most complete victory (against the enemy)' (ibid.: 93). Sun Zi's advice recalls Doenitz's (1959: 425) instructions to his U-boat crews when, following the loss of their bases in western France in 1944, they reverted to operations in the North Sea:

> In these operations in coastal waters it was repeatedly stressed to commanders that they should always act in a manner which would be unexpected by the enemy; for example, when attacking convoys close in to the coast, they attack from shallow waters, inshore rather than from seaward, and also that after an attack, they should similarly try to escape inshore.

In the broader context of naval warfare, the appropriateness of Sun Zi's advice for submarine forces, unconventional in relation to the surface forces for which there was no alternative until the beginning of the twentieth century, is again obvious. Sun Zi considered that the ability to act stealthily is indeed an important attribute of the successful warrior: 'Such is the intricacy of and subtlety of the expert in warfare that he appears to be invisible and without trace. Such is the mystery and myth of the expert in warfare that he is not heard nor detected' (Chow-Hou 2003: 137).

But, according to Sun Zi, 'the strength of an army does not depend on the superiority of numbers' (ibid.: 266). *The Art of War* also teaches that smaller, weaker forces can still take on larger forces and win, and that larger, stronger forces have weaknesses which can be exploited by the less powerful forces. Perhaps like a barrage of anti-ship missiles launched from land, aircraft, surface vessels and submarines simultaneously to overwhelm the US Navy's Aegis defence system, Sun Zi compares the simultaneous use of direct and indirect approaches and forces (*qi zheng*) to 'gushing torrential water [which] tosses stones and pushes boulders because of the force created by its momentum' (ibid.: 116) and to the moment 'when the ferocious strike of an eagle breaks the body of its prey … because of the exact moment and timing of its engagement' (ibid.: 117):

> Thus, the force and momentum of the adept in warfare are so overwhelming and ferocious, and his timing of engagement highly precise and swift.
>
> His stance and (potential) force is like that of a fully-stretched crossbow, and his timing is exact like that of the release of the trigger (of a crossbow) …
>
> Thus, the person adept at warfare creates situations, postures and momentum that resemble rolling boulders falling from the great heights of mountains: this is what is meant by the forces created by a well-commanded army.
>
> (ibid.: 118, 126)

Sun Zi thus anticipated by almost two and a half millennia Clausewitz's famous dictum: 'the best strategy is always *to be very strong*; first in general, and then at the decisive point … there is no higher and simpler law of strategy than that of *keeping one's forces concentrated*' (Clausewitz 1976: 240). By concentrating its forces against the enemy's weak point, by means of precise timing the weaker belligerent can achieve relative superiority and thereby effectively multiply its force. Sun Zi succinctly summarises this advice in a later chapter: 'the disposition and deployment of an army should be to avoid strengths and attack weaknesses' (ibid.: 155).

As Castex explained, the submarine's ability to remain concealed enables it to expand the theatre of operations both spatially and temporally by conferring on it the capacity to be anywhere at any time. This forces the adversary to divide and disperse his forces, thereby effectively multiplying the forces of the side which deploys submarines. The principle involved was described in Sun Zi's *Art of War* in these terms:

> I (use schemes and ploys that) contradict the normal rules of engagement and prevent [the enemy] from reaching his desired destination.
>
> Thus, if I can uncover the dispositions of the enemy while remaining concealed myself, I can keep my forces concentrated and united, and force those of the enemy to be divided and dispersed.
>
> If I can concentrate and unite my entire troop at one place while those of the enemy are scattered at ten different places, then I can use my entire force against one-tenth of his.
>
> Thus I will be the numerically superior and stronger force and he will be the smaller and weaker force.
>
> If I can use a larger and stronger force to attack a smaller and inferior one, those enemies who engage in battles against me will surely be defeated easily.
>
> The enemy must not know the place (battleground) where I intend to attack.
>
> If the enemy does not know where I intend to attack, he must defend in many places.
>
> The more places an enemy defends, the more scattered are his forces, and the weaker is his force at any one point where I am attacking.
>
> (ibid.: 141–3)

Ruses and stratagems

One of Sun Zi's best-known maxims is that which states: 'all war is based on the principle of deception' (ibid.: 22). In accordance with this principle, he advises:

> Thus, when you are capable, feign that you are incapable.
>
> When you are able to deploy your forces, feign that you are unable to do so. When you are near the objective, feign that you are far away; and when you are far away from the objective, feign that you are near.
>
> When the enemy is greedy for small advantages, offer baits to lure him.
>
> (ibid.: 23–4)

To illustrate how Sun Zi's advice might be applied to undersea warfare, Commander Frank Borik, USN, wrote an essay entitled 'Sub Tzu and the Art of Submarine Warfare' which won him first place in the Chairman, JCS,

Strategic Essay Competition in 1995. Borik (1996) creates a fictitious character, Captain Hwei Shwei of the Qingdao Submarine Academy, who writes a paper entitled 'Dragons and Centipedes at Sea: A Strategy for the Modern Offshore Active Defense by the People's Navy', in which he proposes a strategy based on the teachings of Sun Zi and the strategic traditions of China. The proposed strategy relies on a combination of deception and asymmetric stratagems to enable the weaker Chinese naval forces to defeat the stronger US forces. For example, reflecting the potential limitations of the Aegis system, 'Hwei'/Borik suggests that Chinese forces could exploit US forces' reliance on centralised detection, tracking and targeting systems for stand-off weaponry by seeking to confuse and overwhelm the system by the use of large numbers of decoys such as false periscope masts and buoys with radar reflectors. The aim would be to cause US forces to waste their numerically finite anti-submarine weapons on these decoys, noting that during the Falklands War virtually every British anti-submarine weapon was expended on false contacts caused by the environment. 'Hwei'/Borik also suggests that the Chinese could use small fishing or trading vessels to tow acoustic decoys that make the same sounds as submarines to further confuse US ASW systems. The use of carbon-fibre nets by Chinese fishing boats, to catch US submarines, is another asymmetric method of anti-submarine warfare which could be used by Chinese forces.

Goldstein and Murray (2004: 191) also note that Chinese sources openly describe using certain submarines as 'bait', and suggest that the PLA Navy could use its older and less sophisticated submarines to screen higher-value assets or to lure American submarines into revealing their own presence to lurking *Kilos* by executing attacks against 'nuisance *Mings* and *Romeos*'. They speculate that this may well be one reason why China continues to operate vessels which are regarded as obsolete by Western observers.

Littoral warfare: taking advantage of terrain

As we have seen, geography would be an important dimension of any naval war between China and the United States in the context of a Taiwan Strait conflict. A conflict in the narrow seas off the Chinese coast and waters surrounding Taiwan would give Chinese forces all the benefits of coastal navies operating in home waters. It would, for example, enable Chinese naval forces to exploit the advantages of proximity to their bases and ground-based support such as aircraft and coastal missile batteries. Chinese submarine forces would also have the benefit of greater familiarity with the undersea terrain, a knowledge which would be an invaluable asset for tactical manoeuvre. US forces, on the other hand, despite all their efforts to adapt to the exigencies of post-Cold War contingencies, would suffer the disadvantages of a blue-water navy operating in a littoral environment.

Sun Zi's *Art of War* devotes considerable attention to the importance of exploiting terrain for strategic and tactical advantage. Terrain is one of the seven dimensions and five factors that Sun Zi recommends as essential considerations in the strategic planning process:

> Terrain refers to whether the route to be taken is long or short, whether the ground is treacherous or safe, wide or narrow with regard to ease of movement, and whether the ground will determine the death or survival (of an army).
>
> (Chow-Hou 2003: 14)

Victory will go to the side which is better able to exploit weather and terrain (ibid.: 17). For Sun Zi, terrain is an important determinant of force size, composition and structure, since victory depends on an evaluation of the terrain, which enables military planners to estimate the difficulty of the campaign and hence the scope of the operation:

> Based on the assessment of the scope of the operation, the calculation of one's own forces (to be committed to the military campaign) is made. Based on the calculation of one's own forces, comparisons are evaluated against those of the enemy.
>
> (ibid.: 99–100)

Sun Zi thus describes the essential components of the strategic calculation and planning process of relating means and ends that results in the formation of a state's armed forces. As we have seen, geography and balance-of-force calculations have played an essential part in the decision of China's strategic planners to develop a strong undersea warfare capability.

As we have also seen, the currents, bottom composition and topography, varying temperature and salinity layers of the Taiwan Strait present formidable difficulties for ASW operations. Chinese submarine forces are undoubtedly more familiar with the undersea environment of their littoral waters than are US forces. Although Sun Zi clearly had in mind military operations in the terrestrial environment, his warning about the importance of having knowledge and understanding of the physical environment of the operational theatre can be interpreted in a more general sense that is applicable to a naval confrontation in the waters around Taiwan: 'Those who do not know the conditions of the forested mountains, the dangerous terrain of mountain paths, and the treacherous nature of swamps and marshes will not be able to conduct the movement of troops' (ibid.: 182).

Goldstein and Murray (2004: 188) report that PLA Navy submarine commanders are seeking to develop an intimate acquaintance with the topography, thermoclines, currents, and other hydrographic peculiarities of China's coast and particularly in close proximity to Taiwan. They note a recent description of PLA Navy submarine exercises in which submarines stop their engines and either rest on the seabed or drift on a thermal layer as if 'perched in the clouds', and another where submarines practised cloaking as a 'submerged reef' and riding the rapid local currents. Goldstein and Murray (2004: 189) conclude that 'these peculiar phrases suggest that Chinese submariners clearly recognize the importance of having comprehensive local environmental knowledge and exploiting that knowledge to maintain stealth and other tactical advantages'.

Blockade preferable to amphibious assault

There is a particular congruence between Sun Zi's precepts and the most logical strategy – and no doubt the one preferred by PLA strategists – for the use of military force to attain Beijing's political objectives with respect to Taiwan: a maritime blockade of the island of Taiwan by means of mines and submarines. Attacking Taiwan's maritime trade and supply routes not only provides Beijing with a means of attacking Taipei's weak points, but also has the merit of being less costly than a direct attack, whether it be by means of amphibious assault or ballistic missile strikes. A blockade would also have the merit of minimising material damage and casualties on both sides. The PLA and the CCP are still proud of the fact that their forces captured Beiping (Beijing), the once and future capital of China, in January 1949 without a bloody and debilitating fight (Wortzel 2003: 68).

An amphibious assault, even if Chinese forces managed against all probability to seize and hold the island, would inevitably be a bloody affair. As Sun Zi cautioned, the preparation of assault equipment and weapons would be expensive both in time and money, and 'even if he (the general) orders his troops to assault the walls like ants, one-third of them will be killed and the city will still not be conquered' (Chow-Hou 2003: 62).

On the other hand, a blockade, if it succeeded in persuading the Taiwanese to accept Beijing's political conditions, would be a textbook example of Sun Zi's fundamental precept of 'subduing the enemy without any battle'. This well-known maxim is the logical corollary of Sun Zi's over-arching approach which holds that 'in general, when waging war, capturing a whole nation intact is a better strategy; to shatter and ruin is a weaker option' (ibid.: 57). This approach, which seeks to reduce as far as possible the human, material and financial costs of war, explains Sun Zi's renowned hierarchy of preferred strategies, according to which:

The ability to subdue the enemy without any battle is the ultimate reflection of the most supreme strategy.
    The next best strategy is to attack his relationships and alliances with other nations.
    The next best strategy is to attack his army.

> The worst strategy of all is to attack walled cities.
>
> Attack walled cities when there are no other alternatives …
>
> The adept in warfare is able to subdue the army of the enemy without having to resort to battles.
>
> He (the adept in warfare) is able to capture the cities of others without having to launch assaults.
>
> He (the adept in warfare) is able to destroy and damage the states of others without waging protracted campaigns.
>
> He (the adept in warfare) will focus on using effective policies and strategies to keep all his resources intact and yet be able to contest for world supremacy against the other states.
>
> Thus, his troops are not worn out and his victories and gains are complete.
>
> This, in essence, is the art of strategic attacks.
>
> (ibid.: 59–65)

The preference for indirect and economic methods of securing strategic objectives is a leitmotiv of *The Art of War*. Although Sun Zi is always conscious of the need to be prepared to use violent methods and to engage in combat in the absence of any alternative course of action, he always displays a marked preference for methods which avoid the direct and massive use of force. He prefers indirect approaches. He emphasises methods which exploit the weaknesses of adversaries and which operate above all on the opponent's psychology through coercion by means of threats of violence, harassment and attrition, or, alternatively, through positive inducements to cooperate. For example, Sun Zi provides advice on the preferred strategies for subduing or winning over neighbouring warlords. Sun Zi's preferred methods avoid the direct use of massive force, but instead operate primarily on the adversary's psychology by relying on coercion or persuasion through threats of violence, harassment or inducements. To borrow Clausewitz's words (1976: 83), if not his methods, Sun Zi's advice for 'compelling our enemy to do our will' is:

> succumb [sic] the neighbouring warlords through the use of intimidation and threats.
>
> Harass and wear down the neighbouring warlords through incessant creation of troubles and activities.
>
> Hasten and direct the movements of the neighbouring warlords through the offer of benefits and baits.
>
> (Chow-Hou 2003: 224)

This passage of Sun Zi encapsulates the central elements of Beijing's coercive strategy vis-à-vis Taipei. The very existence of a large and – at least so far as its *Kilo*- and *Song*-class submarines are concerned – highly capable

Chinese submarine fleet, able at a moment's notice to place a stranglehold on the trade and communications which represent the lifeblood of the Taiwanese economy, constitutes a threat to Taiwan and other 'neighbouring warlords' who may be inclined to come to Taipei's aid, as do China's short-range ballistic missile batteries across the Taiwan Strait in Fujian. The imposition of a blockade would correspond to Sun Zi's recommended strategy of 'harassing and wearing the neighbouring warlords through incessant creation of troubles and activities'. A blockade would nevertheless represent an indirect strategy, a means of 'subduing the enemy without battle'. It is a method of 'capturing the nation intact' rather than 'destroying and shattering' it. At the same time, Beijing is trying to bend the Taiwanese people to its will by offering inducements in the form of economic links and trade and investment opportunities for Taiwanese businesses on the mainland. Simultaneously, however, Beijing is preparing for the possibility that this indirect strategy of subduing the enemy without fighting may not succeed. The PLA is therefore establishing the means to apply a more direct strategy of direct attack against the enemy's army and 'walled cities' by its batteries of short-range ballistic missiles.

Guerrilla warfare at sea

Classical maritime strategic theory and Chinese strategic culture lend mutual support to the adoption of operational doctrine based on the notion of waging guerrilla warfare at sea. Corbett (1911: 211) had long ago recognised the similarities between guerrilla warfare waged on land and defensive warfare waged at sea by inferior forces disputing control 'by harassing operations, to exercise control at any place or at any moment as we [see] the chance, and to prevent the enemy exercising control in spite of his superiority by continually occupying his attention'. In 1936, Mao Zedong (1963: 109) recalled the fundamental principles of partisan warfare adopted by the Central Committee of the Chinese Communist Party in 1928 in the famous sixteen-character formulation: 'when the enemy advances, we retreat; the enemy camps, we harass; the enemy tires, we attack; the enemy retreats, we pursue.' The concept of waging guerrilla warfare at sea is an ingrained feature of the PLA Navy's strategic culture. One of its earliest and most celebrated victories occurred in January 1955 when PLA Navy PT boats sank one Nationalist warship, while heavily damaging another. The success of this action was due to the PT boats hiding themselves among a concentration of fishing boats and dashing from this cover to surprise the unfortunate Nationalist vessel – a stratagem later celebrated in the Chinese press as an example of guerrilla warfare at sea (Swanson 1982: 189).

Chinese strategic culture, both ancient and modern, although forged almost exclusively from the experience of terrestrial warfare, provides fertile ground for the development of operational concepts for the PLA Navy's submarine fleet based on the concept of waging guerrilla warfare at sea. For Chinese maritime strategists, undersea warfare against the world's most powerful naval force would resemble guerrilla warfare on land not only because of the imbalance of force between the opponents, but also because of the lack of defined lines and fronts which is the characteristic of both maritime and guerrilla conflict.

The stealthiness of submarines would also strike a chord with strategists steeped in Maoist and ancient Chinese military theory. To attack an enemy in such a way that he is unable to discern either the author or the origin of the attack would be consistent with the advice of Sun Zi: 'such is the intricacy and subtlety of the expert in warfare that he appears to be invisible and without trace. Such is the mystery and myth of the expert in warfare that he is not heard or detected' (Chow-Hou 2003: 137). In his 1938 analysis of the strategic problems of the partisan war against Japan, Mao (1963: 158) noted that it was the very weakness of the partisans which enabled them to take the initiative:

> it is precisely because the guerrilla units are small and weak that they can mysteriously appear and disappear behind enemy lines, without the enemy's being able to do anything about them, and can thus enjoy a freedom of action such as massive regular armies never can.

Deserts and seas

China's long military tradition of defending its periphery against nomadic raiders from the deserts and steppes presents certain structural similarities to the PRC's current preoccupation with defending its maritime periphery from seaborne attack. China's northern and north-western borders cross relatively open and flat grass and scrublands, deserts and dry steppes (Swaine and Tellis 2000: 9). For most of China's history as a unified state – from the period of the consolidation of the feudal kingdoms and fiefdoms under the Qin dynasty (221–206 BC) – the northern and north-western land frontiers were the central front for the defence of the Chinese Empire.

Historically, the deserts and steppes of China's northern and north-western frontier regions have served as a zone of separation and confrontation between the settled agricultural societies of China and the nomadic, pastoral peoples of the steppes. These peoples posed a military threat to the sedentary Han Chinese because of their superior war-fighting capabilities and high mobility. They were expert horsemen and skilled at using the bow and the sword to conduct lightning raids against China's usually static defences. They were also usually able to evade pursuit by China's larger, but slower, infantry-based and heavily armoured military forces by melting back into the vast expanses of the steppes and deserts (ibid.: 29). In his consideration of the influence of Inner Asia on China's military history, John Fairbank (1974: 11) observed that:

> the rough equivalent of the Mediterranean Sea in ancient China's military experience was the limitless expanse of the grasslands and deserts of Inner Asia, where China's intensive agriculture could never be established and the menace of raiding nomad cavalry could never quite be destroyed. Caravan expeditions into Inner Asia, lacking ships, of course faced enormous supply problems. The fertile oases of Kashgaria provided ports of call and even bases for task forces on the trade routes, while raiding parties could navigate far across the Mongolian grasslands.

The observation that deserts and seas have an analogous strategic functional similarity is not new. Commenting on the strategic rivalry between the Romans and the Carthaginians, Mahan (1965: 20–1) drew an analogy between warfare in a desert environment and warfare in a maritime environment. T. E. Lawrence (1988: 264), planning his operations in support of the British invasion of Syria in 1917, conceived of his irregular camel-mounted raiding parties against Turkish forces as a kind of

> naval war, in mobility, ubiquity, independence of bases and communications, ignoring of fixed ground features, of strategic areas, of fixed directions, of fixed points. 'He who commands the sea is at great liberty, and may take as much or as little of the war as he will.' And we commanded the desert. Camel raiding parties, self-contained like ships, might cruise confidently along the cultivation-frontier, sure of an unhindered retreat into their desert-element which the Turks could not explore.

Thus, from the strategist's perspective, deserts and seas present certain similarities. Both are unpopulated or sparsely populated expanses, where isolated pockets of inhabited terrain – oases or islands – assume a strategic importance. Both are traversed by communications routes which represent the most direct practical route from point A to point B. The standard responses to threats to these lines of communication from brigands or pirates are similar: organised protection by caravans or convoys. Like the sea, the desert is a zone of separation and often confrontation between distinct geopolitical areas (Chauprade 2001: 165).

Strategically, a major challenge for generations of Chinese military leaders was to secure China's northern and western land frontiers and inland trade routes across Central Asia to the Middle East and beyond – the Silk Road – against the threat of 'barbarian' raids and invasions. Just as the need to protect vulnerable oceanic trade routes was a key motivating factor in the expansion of the British Empire and the Royal Navy, the protection of trade routes through Central Asia was a major driving force behind the expansion of Imperial China, particularly during the ascendant phases of the Han, Tang, Ming and Qing dynasties. The challenge for imperial Chinese military forces of securing trade routes and protecting the frontiers of its settled, agricultural lands against nomadic, horse-borne raiders and invaders resembles structurally the contemporary challenge of defending China's southern and eastern maritime frontiers against the threat of seaborne and airborne raiders and invaders. David Wright (2002: 57) cites the specific strategic recommendations for dealing with the threats posed by the nomads contained in the *New Tang History* (*Xin Tangshu*), a work largely written and edited by the Song dynasty Confucian scholar Ouyang Xiu (1007–72):

> Our Chinese infantrymen are at their best in obstructing strategic passes, while the barbaric cavalrymen are at their best on the flat-lands. Let us resolutely stand on guard [at the strategic passes] and not dash off in pursuit of them or strive to chase them off. If they come, we should block strategic passes so that they cannot enter; if they withdraw, we should close strategic passes so that they cannot return.

Those regimes which adopted a military capability to take the fight to the enemy rather than rely on static defences were the most successful in subduing the nomadic and semi-nomadic tribes and pacifying the periphery. The Mongol Yuan and Manchu Qing dynasties maintained large cavalry units which enabled them to conduct offensive, mobile operations into the steppes, deserts and high plateaux of China's terrestrial periphery (Wright 2002: 60). During the ascendant phase of the Ming dynasty under the first two major regencies – that of Hong Wu (1368–99) and Yong Le (1403–23) – a similar offensive strategy pushed the Ming's boundary of control far to the north into the Mongolian steppe lands (Johnston 1995: 184). But as Ming military power and offensive capabilities declined relative to that of the Mongols, Ming security lines were pulled back and, particularly after a disastrous defeat at Tu Mu in July 1449 where the emperor was captured and the Mongols reached the walls of Beijing, Ming strategists adopted a more defensive posture, based on static defensive measures to control strategic sites and choke points such as wall-building, the establishment of a better system of warning towers and beacons, the stockpiling of rations and supplies, and improvements in training and discipline (ibid.: 179). This defensive posture did not preclude the use of mobile troops to launch attacks beyond the border to pre-empt and intercept Mongol raiding parties, but these offensive operations were limited in scope and duration.

There are echoes of the strategies adopted for the defence of Imperial China's northern land frontiers against the incursions of the barbarian nomads in the PRC's plan to develop a blue-water navy, spearheaded by a powerful submarine force, to defend contemporary China's eastern and southern maritime periphery. Like the defensive strategy of the Tang and Ming infantrymen guarding the strategic passes against the more powerful and mobile Xiong-nu and Mongol forces, the defence of the PRC's maritime periphery against the more powerful US forces relies on a combination of relatively static land-based air and missile batteries, mines and forward defence by more mobile naval surface and submarine forces. You Ji (1999: 191) has noted, for example, that China's submarine units 'will be able to form an ambush platform at the strategic chokepoints in the West Pacific such as around the Bashi Channel and the Taiwan Strait'. Or, like the offensive strategy of the Yuan and Qing dynasty cavalry units, 'with the growing strength of the 09 nuclear submarine unit and modernised conventional attack submarines, the Chinese will be able to provide a second echelon to support the "wolf pack" of patrol submarines'. The build-up of mobile, short-range ballistic missiles in coastal regions adjacent to the Taiwan Strait, and the 1996 missile tests into waters adjacent to Taiwan's principal ports, recalls the Ming tactic of flaunting its military capabilities, and raising lots of banners and flags and firing off signal cannons to intimidate and unsettle the attacking Mongols.

Castex's description of the operational advantages that submarines enjoy over surface combatants recalls the contrast between the operational role for the mobile and independent Ming cavalry units, and that of the less mobile and more vulnerable heavy infantry units, stationed in relative security of the walls, fortresses and strategic passes to defend the northern frontiers against the Mongol invaders. Unlike surface combatants (or heavy infantry units), submarines are more autonomous (like light, mobile cavalry units), and, according to Castex, do not have to be concerned about being isolated and can be left to their own devices. Moreover, submarines are less bound by the constraints of resupply than surface warships. In addition, their greater range enables them to undertake long patrols and significantly increases their area of operation:

> Many actions and zones which are not possible for surface ships are within the scope of the submarine which has no reason to concentrate systematically in close formations to occupy only a small area of the theatre of operations. It can go on its own [aller en cavalier seul] almost anywhere.
>
> (Castex 1997: vol. I, 13)

Surprise and pre-emption

Classical Chinese and Maoist strategic theory, Western maritime strategic theory and analyses of historical and contemporary Chinese strategic behaviour also tend to support the view that in any conflict involving the use of force against a stronger opponent, Chinese naval warfare doctrine is likely to emphasise the use of swift and pre-emptive surprise attacks to seize the strategic initiative and to administer a psychological or political shock to the enemy.

Sun Zi's advice was to 'attack the enemy when he is not prepared' and to 'move, appear and strike at areas where the enemy least expects you' (Chow-Hou 2003: 28). Wu Qi (440–361 BC) – known by later generations as Wu Zi – the famous general of the state of Wei whose military thought, along with that of Sun Zi, constitutes one of the Seven Military Classics of ancient China, is more elaborate than Sun Zi in counselling the use of surprise and the exploitation of the enemy's weaknesses. The tactics of surprise and pre-emption are always considered as a means of concentrating one's own forces against the enemy's weak points by choosing the optimum timing in order to achieve a relative superiority:

> In employing the army you must ascertain the enemy's voids and strengths and then race [to take advantage of] his endangered points. When the enemy has just arrived from afar and their battle formations are not properly deployed, they can be attacked. If they have eaten but not yet established their encampment, they can be attacked. If they are running about wildly, they can be attacked. If they have laboured hard, they can be attacked. If they have not yet taken advantage of the terrain, they can be attacked. When they have lost the critical moment and not followed up on opportunities, they can be attacked. When they have traversed a great distance and the rear guard has not yet had time to rest, they can be attacked. ... In general circumstances such as these, select crack troops to rush on them, divide your remaining troops, and continue the assault – pressing the attack swiftly and decisively.
>
> (Sawyer 1993: 213)

Sun Zi also counselled the initial adoption of a strong defensive posture, making use of stealth and concealment, as a prelude to offensive operations the effectiveness of which would be enhanced by the use of unpredictability:

> Sun Zi said: In ancient times, those who were skilful in warfare ensured that they would not be defeated and then waited for opportunities to defeat the enemy ... The adept in defence is able to conceal his forces in the most secretive ways and places of the earth.
>
> (Chow-Hou 2003: 88–92)

But the classical Chinese military prescription for action in situations where concealment has failed is also to launch a surprise attack. In T'ai

Kung's *Six Secret Teachings* – which purportedly record the political and tactical advice of the famous soldier, official and sage T'ai Kung to Kings Wen and Wu of the Chou dynasty – King Wu asks: 'If the enemy knows my true situation and has penetrated my plans, what should I do?' T'ai Kung replies: 'The technique for military conquest is to carefully investigate the enemy's intentions and quickly take advantage of them, launching a sudden attack where unexpected' (Sawyer 1993: 52).

The study of Chinese strategic culture points to the existence of certain features which could incline Chinese decision-makers towards the adoption of a surprise pre-emptive strike. The study of naval strategic theory and the use of the submarine as an asymmetric instrument of naval warfare – 'the ideal tactical weapon of the offence', according to Doenitz, for what Castex called strategies of 'offensive defence' – suggests that the existence of a powerful submarine arm in the PLA Navy's fleet would reinforce the features of Chinese strategic culture with a logic of appropriateness favouring the adoption of such a strategy. Finally, the study of the strategic behaviour of the PRC since its foundation appears to show that Beijing has not been as reluctant to use force to achieve its strategic objectives as conventional interpretations of Chinese culture might suggest.

11

CHINESE STRATEGIC CULTURE – SUBMARINES AND PROSPECTS FOR WAR IN THE TAIWAN STRAIT

Chinese realpolitik

The analysis of Chinese strategic behaviour at the grand strategic level provides evidence of the operation of a strategic culture which predisposes Chinese decision-makers towards the pre-emptive, offensive use of force in response to perceived threats to China's national security. Johnston's detailed analysis of the Seven Military Classics and Ming dynasty operational strategy against the Mongol threat on the northern frontier reveals the existence of a strategic culture which accepts that 'warfare and conflict are relatively constant features of interstate affairs, that conflict with the enemy tends towards zero-sum stakes, and consequently that violence is a highly efficacious means for dealing with conflict' (Johnston 1995: 61). Johnston labels (ibid.: 61) this view of security as a *parabellum* (a term derived from the maxim *si pacem parabellum* – 'if you want peace, prepare for war') or realpolitik paradigm 'whereby the sine qua non of state security is sufficient military capabilities and, preferably, the military defeat of the adversary'. The influence on strategic behaviour of this *parabellum* paradigm is tempered by an appreciation of the need for a pragmatic response to security threats embodied in the notion of *quan bian*, or 'absolute flexibility' in the application of force – the idea that:

> given that constant change is the key characteristic of conflict situations, a strategist must be prepared to adapt to dangers and opportunities as they suddenly appear. The strategist cannot be restricted, constrained by or wedded to self-imposed a priori political, military or moral limits on strategic choices.
>
> (ibid.: 102)

Johnston's conclusions about the existence of a Chinese strategic culture characterised by a dominant *parabellum* paradigm mediated by the concept of *quan bian*, contrast with the accepted view among many Western and Chinese analysts of traditional and contemporary Chinese strategic behaviour

149

that Chinese strategic thought 'disesteems' the use of violence to resolve conflicts. Johnston's analysis (ibid.: 155) of the Seven Military Classics leads him to conclude that in fact two distinct strategic cultures coexist:

> One, derived from a broad Confucian–Mencian central paradigm, places non-violent, accommodationist grand strategies before violent defensive or offensive ones in a ranking of strategic choices … The other, derived from the *parabellum* paradigm, generally places violent, offensive strategies before static defence and accommodationist strategies.

Johnston suggests that the Confucian–Mencian strand of Chinese strategic culture serves an essentially symbolic function in justifying the use of violent means to attain political ends, while the *parabellum* element provides the dominant operational paradigm for strategic action. Johnston (ibid.: 216) concludes that, in accordance with the dominant influence of this *parabellum* paradigm in Chinese strategic culture:

> Ming decision-makers preferred, in principle, more offensive uses of force (including both external extermination campaigns and active defence measures) to static defence and accommodation. When the latter two were advocated – as opposed to being preferred in principle – it was generally on the grounds that temporally bounded strategic conditions (i.e. the present balance of capabilities) undermined the efficacy of offensive strategies. In other words, static defence and accommodation were strategies of last resort … One also finds evidence that Ming strategic choice reflected this contingent *parabellum* calculus. In other words, Ming China tended to adopt more coercive, offensive strategies towards the Mongols when it was capable of doing so. Ming decision-makers resorted to defensive or accommodationist strategies when their ability to mobilize resources for offence was more constrained.

While acknowledging that there is some debate about whether traditional Chinese patterns of thought and interaction have carried over into the post-1949 period, Johnston (ibid.: 256) nevertheless suggests that at the level of aggregate behaviour there is some evidence of the continuing influence of *parabellum* strategic culture on contemporary Chinese security policy. He observes that there is a great deal of consistency between much of Mao Zedong's strategic thinking from the 1930s through to the 1950s and that of traditional Chinese strategic thought. It is therefore not surprising that at the level of aggregate behaviour there is evidence of continuing *parabellum* strategic culture on contemporary Chinese strategic policy – especially given the fact that Mao exerted a virtual monopoly over the PRC's strategic thinking in the post-1949 period and that his successors still explicitly look to Mao's strategic thinking to validate their own. He cites (ibid.: 256) evidence from the analysis of foreign policy crises by Wilkenfeld, Brecher and Moser which shows that of the 11 foreign policy crises in which the PRC had been involved prior to 1985, it had used violence in eight (72 per cent) of them – proportionally more than the other major powers in the twentieth century: the comparable figures for the US, the USSR and the UK from 1927 to 1985 were respectively 18 per cent, 27 per cent and 12 per cent.

Other analysts of Chinese strategic behaviour support Johnston's conclusions about a disposition among Chinese decision-makers to regard force as an effective means to attain national security goals. Andrew Scobell (2003: 5), for example, also contends that Chinese strategic culture is composed of two strands, both shaped by an ancient and enduring civilisation: 'a distinctly Chinese pacifist and defensive-minded strand, and a Realpolitik strand favouring military solutions and offensive action.' Scobell (ibid.: 15) argues that both of these strands are active and they influence and combine in a dialectic fashion to produce what he calls a 'Chinese Cult of Defence'. The effect of this 'Cult of Defence' on Chinese strategic behaviour is, paradoxically, 'not a preference for what are clearly defensive military policies and actions but rather those that are actually offensive' (ibid.: 26). Like the 'Cult of the Offence' which played a significant role in precipitating the First World War, the Chinese 'Cult of Defence', according to Scobell (ibid.: 26), glorifies flawed strategic assumptions and increases the likelihood of the resort to war.

Swaine and Tellis (2000: 44) also conclude that the historical record of Chinese strategic behaviour suggests a propensity to resort to war:

> A cursory examination of the security behaviour of the Chinese state suggests that Chinese rulers have frequently resorted to violence to attain their national security objectives. In fact, one could argue that the use of force has been endemic in Chinese history. According to one Chinese military source, China engaged in a total of 3,790 recorded internal and external historical wars from 1100 BC (Western Zhou) to 1911 (end of the Qing Dynasty). These included both violent internal conflicts during periods of internal division and conflicts with non-Chinese powers. Moreover, in the Ming alone, China engaged in an average of 1.12 *external* wars per year through the entire dynasty.

Swaine and Tellis (2000: 65) also agree with Johnston's conclusion that the inclination in Chinese strategic culture to consider the use of force as a highly efficacious means to achieve security objectives is tempered by the more pragmatic calculations of relative material capabilities. They conclude that:

The historical record suggests that the Chinese state has frequently employed force against foreign powers but generally followed a pragmatic and limited approach to the use of such force. Specifically, it has employed force against foreigners primarily to influence, control, or pacify its strategic periphery and generally has done so when it possessed relative superiority over its potential adversaries on the periphery ... However, an inability to establish a material position of superiority over the periphery through military force – or strong levels of domestic opposition to the use of such force – often led to the adoption by the state of non-coercive methods, usually involving appeasement and passive defences, which frequently provided long periods of security from attack.

If Johnston's conclusions are correct, it is reasonable to suppose that contemporary Chinese strategic culture is characterised by a similar duality to that which characterised Imperial China's strategic culture: a preference for more defensive and accommodationist strategies when the balance of strength between China and her adversaries was unfavourable (from the Chinese perspective), and for more aggressive, offensive strategies when the balance of strength was in China's favour. If this is so, then Chinese strategic behaviour would appear to be a particular instance of the offence–defence theory which contends that international conflict and war are more likely when offence has the advantage, while peace and cooperation are more probable when defence has the advantage (Lynn-Jones 1995: 661). It would therefore follow that the Pentagon has some justification in considering that the risk of Beijing resorting to force to try to resolve the Taiwan issue is growing with the modernisation and transformation of the PRC's military capabilities. This risk grows all the greater as the development of Taiwan's military capabilities fails to keep pace with those of the PLA and the gap between the military capabilities on either side of the Taiwan Strait continues to widen.

As we have seen, however, the current balance of force between the two sides of the Taiwan Strait favours the mainland in two dimensions of potential conflict in particular: ballistic missiles and submarine forces. A combination of passive and active defence measures could probably limit the potential risk posed by the former threat. However, without an effective tactical submarine capability of its own, the anti-submarine warfare capabilities of the Taiwanese military forces are unlikely to be able to counter effectively an undersea warfare campaign conducted by the PLA Navy. Even in the case of an American intervention, US military forces would have difficulty in dealing successfully with Chinese submarines in the difficult conditions of littoral warfare in the seas surrounding Taiwan. It therefore would not be unreasonable for the PLA Navy to aim for sufficient superiority in undersea warfare to enable it to achieve at least the control of the subsurface of the Taiwan Strait.

Despite this potential local superiority in undersea warfare capability, it is unlikely that the PRC would start a war with Taiwan without having some confidence that its military forces would be able to prevail in other dimensions of the conflict. However, in the context of a crisis where the political and military authorities in Beijing faced politically unacceptable alternatives, might they not be tempted to use their best weapon in an attempt to get out of an impasse? In such a situation, the idea of a surprise, pre-emptive attack against US carrier battle groups by the PLA Navy's tactical submarines cannot be dismissed, particularly as the logic of such an action would be reinforced by elements of Chinese strategic culture, inclining Beijing's decision-makers to adopt such a desperate course of action.

The logic of asymmetric warfare

As we have seen, a prominent, enduring and pervasive feature of Chinese strategic culture, both ancient and modern, is the preoccupation with strategies and tactics to enable the weak to overcome the strong. The importance of asymmetric warfare in Chinese military thought is arguably one of the features which distinguishes it from Western military tradition. Paul Godwin (2003a: 23) sees the concern to compensate for deficiencies in arms and equipment as one of the central sources of continuity in the development of doctrine, strategy and concepts of operations of the PLA. As a consequence, 'China's defence strategies and policies have focused primarily on preparing the PLA for possible confrontation with vastly superior adversaries.'

The strategic logic of asymmetric warfare which places a premium on pre-emption and surprise to achieve the psychological impact on the adversary is reinforced by the disposition towards the pre-emptive use of force which is inherent in China's strategic culture and behaviour. This is because, unlike conventional warfare which is aimed at material objectives (the conquest of territory or the destruction of the enemy's forces), the objectives of asymmetric warfare are often immaterial. As Jacques Baud (2003: 101) observes:

> Asymmetry uses a logic which results from a systematic analysis of the conflict and of the adversary, his decision mechanisms and the relationship between the decision-makers and the society as a whole. Its effects are measured more for their impact on decision processes than for their impact on the forces physically engaged in the conflict. Western societies, with their often superficial reading of conflicts, are particularly vulnerable to asymmetric approaches to warfare.

Asymmetric warfare strategies also depend for their success on an asymmetry of interests at stake. The psychological impact of the use of surprise and extreme violence is likely to have greater deterrent effect when the victim has less than vital interests at stake in the conflict. Chinese strategic analysts writing in military journals frequently cite the United States' experiences in Lebanon in 1982 and Somalia in 1993 to illustrate this point (Burles and Shulsky 2000: 75). It is a common belief among Chinese strategists that the United States could not possibly have a greater interest in Taiwan than China has. A widely cited expression of this view is the implicit nuclear threat made during a visit to the United States in October 1995 by the Deputy Chief of PLA General Staff, General Xiong, in a conversation with a senior Pentagon representative, US Assistant Secretary of Defense, Chas Freeman. Xiong is reported to have said that Washington cared more about Los Angeles than it did about Taiwan and would therefore be deterred from intervening in a mainland–Taiwan war by the risk of a Chinese nuclear strike (Scobell 2003:186).

The perception among China's policy elite that the interests of China and the United States in the Taiwan Strait are asymmetric may lead Chinese decision-makers to believe that China's limited nuclear deterrent would deter the United States from the use of conventional force to defend Taiwan. Alternatively, they may believe that this deterrent would provide Beijing with greater freedom of action to conduct pre-emptive conventional strikes against US forces or facilities in the event that it failed to deter such US intervention. However, at least until China's new DF31 solid fuel, road-mobile intercontinental missile becomes operational – probably in 2007 – China's leaders could not be confident that their current limited arsenal of cave- and silo-based liquid-fuelled intercontinental and intermediate-range nuclear-armed missiles constitutes a sufficiently reliable second-strike force to deter the United States from intervening in a cross-strait conflict. And even then, as we have argued in Chapter 2, with the deployment by the US of theatre and strategic ballistic missile defence systems and advances in the ability of the United States to use space-based assets to locate and destroy moving targets, the Chinese leadership's confidence in the reliability of their land-based nuclear deterrent forces is set to decline further. Uncertainty about the deterrent effectiveness of their nuclear weapons would be reinforced by Chinese strategists' analysis of occasions in the past where nuclear weapons have failed to deter aggression against their possessors. Egypt's 1973 attack against Israel, Argentina's invasion of the Falklands/Malvinas in 1982, and Iraq's invasion of Kuwait in 1990 and its ballistic missile attacks against Israel are the classic cases of nuclear deterrence failure. But Chinese leaders will be even more acutely aware that China itself has on several occasions demonstrated its willingness to risk aggression against nuclear states or the allies of nuclear states – thereby providing practical confirmation of Mao Zedong's judgement that nuclear weapons are in fact 'paper tigers' (Lewis and Xue 1988: 6).

It is more likely that a belief in asymmetry of Chinese and American interests in the issue of Taiwan's sovereignty would reinforce Chinese leaders' appreciation of the need for an effective conventional deterrent to American intervention in a Taiwan Strait conflict – such as a strong and effective tactical submarine force. Thus China's evolving submarine force is set to play an increasingly important role in the trilateral deterrence relationship which structures strategic interaction in the Taiwan Strait. This triple deterrence dynamic is composed of the measures taken by Beijing to deter Taipei from declaring Taiwan's independence from China; Washington's efforts to deter Beijing from resorting to armed force to secure the reunification of Taiwan with China; and Beijing's attempt, in response to American threats, to deter Washington from coming to Taipei's assistance in the event of a crisis or conflict. The *Kilo* submarines, the *Sovremenny* destroyers, the SU-30MKK fighter-bombers, sophisticated sea-mines, ballistic missiles, shore-based cruise missiles and, ultimately, China's nuclear forces all contribute to a system of deterrence designed to raise the potential costs of American intervention to a level where they would outweigh the value of American interests in preserving Taiwan's autonomy.

Once this conventional capability is obtained, this belief may in turn exaggerate Chinese military and political leaders' assessment of its effectiveness as a deterrent. Optimism about the efficacy of their conventional deterrent, whether justified or not, may reinforce the inclination towards the pre-emptive use of force that is a feature of Chinese strategic culture.

The quest for a conventional deterrent, an asymmetric capability which would provide Chinese forces, despite their overall relative weakness, with a means to inflict a strategically decisive blow against the superior forces of the United States, is a prominent theme in the writings of Chinese strategic analysts. Most often, this 'trump card' or the 'killer mace' (*shashoujian* or *sashoujian*) (Ross 2002: 72) is seen as a capability to undermine US information dominance and electronic warfare superiority, but it also encompasses any highly advanced technology which could give Chinese forces the ability to exploit a chink in the US armour.

The likelihood of the operational strategy of China's submarine force reflecting its patterns of strategic behaviour at higher levels is something that we can only infer from the analysis of the logic of maritime strategic theory. Since the PLA Navy, unlike the US Navy, does not publish information on its operational doctrine, there is no direct, open-source evidence available about how Chinese submarine forces would operate in wartime. Do the patterns of China's use of force at the grand strategic level indicate that these same patterns apply at the levels of military strategy, operational doctrine and tactics? China's principal putative adversary certainly believes that this is likely to be the case. In its 2002 report to Congress on the military power of the PRC, the Pentagon (USDoD 2002: 13–14) notes that during the 1990s a shift took place in the PLA's operational theory from predominantly annihilative to coercive war-fighting strategies:

> Shock and surprise are considered by PLA strategists as crucial to successful coercion. Accordingly, PLA operational theory emphasizes achieving surprise and accruing 'shock power' during the opening phase of a campaign ... Throughout the 1990s, PLA writers have highlighted pre-emptive strikes as a means to offset advantages that a technologically superior power brings to the fight ... PLA writers have asserted that offensive strike assets, which are more cost effective than defensive assets, must focus on an opponent's ability to conduct strikes and/or conduct counterattack operations ... With no apparent political prohibitions against pre-emption, the PLA requires shock as a force multiplier, to catch Taiwan or another potential adversary such as the United States, unprepared. Observers such as PLAAF Chief of Staff LTG Zheng Shenxia have noted that without adopting a pre-emptive doctrine, the chances of a PLA victory are limited.

Given the evidence that the PLA Navy's doctrine places some emphasis on surprise pre-emptive attacks as a means of compensating for its weakness relative to that of potential adversaries, there is a strong possibility that one of the first actions in a war between the PRC and the United States would be a surprise attack by Chinese submarines against units of an American carrier battle group.

This raises the issue of whether the mere possession of a large and capable tactical submarine force by the PLA Navy could in itself be a leading causal factor in a future war between the PRC and the United States. Offence–defence theory holds that military technology can favour the aggressor or the defender, and that technologies which make it easier for offensive strategies to succeed also make war more likely (Van Evera 1999: 160). Is the tactical submarine a weapons system the possession of which could amplify the propensity for surprise pre-emptive attacks so prominent in Chinese strategic culture and increase the risk of a crisis in the Taiwan Strait degenerating into a Sino-American war? In other words, is there in the tactical submarine something that, in Thomas Schelling's (1966: 234) words, 'we might call the "inherent propensity toward peace or war" embodied in the weaponry, the geography, and the military organization of the time'?

It is certainly true that since it first became a viable weapon system, and contrary to initial expectations that it would play a predominantly defensive role, the submarine has been used frequently, if not predominantly, operationally and tactically as an offensive weapon. Moreover, the submarine has a history of use in surprise attacks, confirming Robert Jervis' (1977: 49) observation that 'weapons and strategies that depend for their effectiveness on surprise are almost always offensive'. And the submarine has a history of being used as a first-strike weapon system in the initial hours of conflicts.

At the beginning of the First World War, for example, one of the first engagements between German and British naval units occurred on 10 August 1914, six days after the two countries entered the war. On that day, the *U15* surprised three British dreadnought battleships halfway between Orkney and Shetland and fired a torpedo at HMS *Monarch*. The torpedo missed its target, but the unexpected appearance of a German submarine so far from its base and so close to theirs struck the Royal Navy with consternation (Van der Vat 1995: 50). On the first day of the Second World War, 3 September 1939, another German submarine, the *U30*, launched a surprise attack against the British liner SS *Athenia* in the Atlantic Ocean off the north-west coast of Ireland (Van der Vat 1995: 162). In the Pacific War, Japanese plans for the attack on Pearl Harbor on 7 December 1941 gave a major role to submarines. According to Hezlet (1967: 192), 27 submarines were disposed around the Hawaiian Islands. Their purpose was to launch five midget submarines to attack the American fleet in harbour at the same time as the carrier-borne aircraft. In fact, according to Hezlet (ibid.):

> The Japanese expected more from the submarines than from the carrier-borne air attack itself. In this they were to be disappointed: the midget submarines achieved nothing; only one managing to penetrate the harbour and all of them being destroyed by the defences.

Immediately following the attack on Pearl Harbor, US Navy submarines were instructed to conduct unrestricted warfare against Japanese ships. The first United States Navy vessel to go on the attack was the submarine *Gudgeon* which left Pearl Harbor on 11 December 1941 to patrol the Japanese Inland Sea, where it became the first American submarine to sink an enemy warship when it sank a Japanese submarine on 24 January 1942 (Van der Vat 1995: 245). One of the first naval actions in the Falklands War was the torpedoing of the Argentine cruiser the *General Belgrano* by the British nuclear attack submarine, HMS *Conqueror*, on 1 May 1982 (Van der Vat 1995: 338). Thus the submarine has something of a history of being the chosen instrument for striking initial, surprise blows at the outset of conflicts.

Despite the submarine's history of use as an instrument of choice for executing surprise attacks, it is important to bear in mind Colin Gray's (1992a: 28) salutary reminder that weapons are not inherently offensive or defensive. Their character (e.g. stabilising/destabilising, offensive/defensive) is determined by the purposes to which they are put and the context in which they are used, rather than by any quality that is intrinsic to them.

Weapons are merely instruments in the hands of their users – means to tactical, operational, strategic and, ultimately, political ends. As Gray (ibid.: 26) points out, it is governments, not weapons, that make war. Nonetheless, weapons do have an influence not only on *how* a war is waged, but also on *whether* it is waged. As Thomas Schelling (1966: 234) remarks:

> Arms and military organizations can hardly be considered the exclusively determining factors in international conflict, but neither can they be considered neutral. The weaponry does affect the outlook for war and peace. For good or ill the weaponry can determine the calculations, the expectations, the decisions, the character of a crisis, the evaluation of danger and the very processes by which war gets under way. The character of weapons at any given time determines, or helps to determine, whether the prudent thing in a crisis is to launch war or to wait.

Weapons exert this influence on the outlook for war and peace not necessarily through any inherent quality they may possess so much as through the calculations, the expectations and the decisions of those who wield them. Bernard Brodie put his finger on the way in which weapons exert their influence on warfare when he wrote that

> there is nothing automatic about the influence of weaponry on warfare. That influence has to be exerted initially through the minds of men, who make judgements, first, about the utility of weaponry or other devices, and, second, about the tactical and strategic implications of the general adoption of these weapons or devices.
>
> (cited by Lynn-Jones 1995: 677)

The fact that the submarine has proved itself in the past to be the 'ideal tactical weapon of offence' for strategies of 'offensive defence' influences judgements about its utility as an instrument of future strategic, operational or tactical purpose. Above all, perceptions by PLA Navy strategists that the submarine has proved its worth as an offensive weapon particularly suited to surprise, pre-emptive attacks against superior enemy forces, may lead them to believe that a large and effective submarine force could give China what Stephen Van Evera calls 'a first strike advantage' in a future conflict with the United States.

A first strike advantage, according to Van Evera's (1999: 37) definition, 'obtains when an advantage accrues to the first of two adversaries to use force'. A first strike advantage has similar effects to a 'first mobilisation advantage', which 'obtains when an advantage accrues to the first of two adversaries to mobilize its forces or otherwise prepare for war'. First strike and first mobilisation advantages, to which Van Evera applies the common label of 'first move advantage' – whether these advantages are real or perceived – can be important catalysts for war. As Van Evera (1999: 37) explains:

> A first-strike advantage creates a direct temptation to use force (for example, Israel's 1967 attack on Egypt). A first mobilization advantage causes war indirectly, by tempting states to mobilize their military forces (for example, Russia's 1914 military mobilization). Such mobilization can trigger war through its secondary effects – specifically, through the preventive or pre-emptive incentives to attack that it can create.

Thus, given that submarine forces are widely perceived to possess inherently offensive characteristics, the very existence of the PLA Navy's large and increasingly effective undersea warfare capability could foster perceptions among the PRC's military and political leadership that this capability could offer China a first move advantage over the United States' military forces in a future Taiwan Strait crisis. The PLA Navy's large and increasingly effective submarine arm could therefore serve to amplify the propensity for pre-emptive action inherent in Chinese strategic culture. And the more effective the PLA Navy's undersea warfare capability becomes with the addition of new advanced boats of the *Kilo* and *Song* classes, the more difficult it will become to manage future crises between Beijing and Taipei so as to avoid a slide into war.

The possession of an effective offensive weapon system, together with the proven willingness of the PRC military and political leadership to take risks and to use force against militarily superior adversaries in the past, and the risk that they might do so again in a future Taiwan Strait crisis, fits a pattern of asymmetric war initiation by weaker powers identified by T. V. Paul. On the basis of comparative case studies of six wars initiated by weaker powers against more powerful opponents, Paul (1994: 30) found that the possession of an offensive weapon capability was a key factor in the decision by weaker states to initiate a conflict with a stronger adversary, especially when such a capability offered the prospect of quick success by striking first. For example, the first important war of the twentieth century, the Russo-Japanese War, began with Japan's surprise torpedo attack on the Russian Fleet at Port Arthur in February 1904, using its newly acquired, technologically advanced Whitehead torpedoes (ibid.: 53). Japan's short-term advantage over the United States in aircraft carriers, high-quality aircraft and air-launched torpedoes in 1941 was also an important factor in the recommendation by Japanese naval strategists to strike first in order to destroy the US aircraft carriers in Pearl Harbor (ibid.: 71).

A second factor identified by Paul which predisposes weaker states to initiate asymmetric wars is the political and military leadership's optimism in their ability successfully to implement limited aims/fait accompli strategies. Paul (ibid.: 24) found that:

> The possibility of asymmetric war initiation is high if the weaker state's decision-makers believe in the efficacy of a successful limited aims/fait accompli strategy. The expectation that a limited aims/fait accompli strategy would help to attain the objectives at stake gives the weaker state incentive to strike, as decision-makers may fear that with the passage of time such a strategy would fail to produce successful outcomes.

With a limited aims/fait accompli strategy, the weaker state, rather than seeking a total victory or the unconditional surrender of the opposing forces, aims only to secure sufficient advantage over the opponent to strengthen its hand in a subsequent bargaining process. Weaker powers typically count on superior strategy and tactics to compensate for inferior military strength. A dominant feature in limited aims/fait accompli strategies of weaker powers, according to Paul (ibid.: 26), are offensive–defensive military doctrines which 'presuppose quick offensive military thrusts followed by a defensive posture to create a fait accompli situation in order to preserve the limited gains until political settlements can be achieved'.

Decisions by weaker states to initiate asymmetric conflicts also tend to be based on the belief that factors such as weapons capability, rapid action and greater resolve to win can create short-lived and geographically circumscribed opportunities for the weaker forces to exploit temporary weaknesses and vulnerabilities in the stronger side's defensive posture. Surprise, according to Paul (ibid.: 29), is a key element of a limited aims/fait accompli strategy. Surprise attacks can destroy a good proportion of the enemy's forces and create a temporary superiority in capabilities and firepower for the weaker power:

> Therefore, in most instances, surprise has been an instrument in the hands of the weaker state, providing the initiator with an advantage that it may not otherwise enjoy. Surprise can sometimes not only work as a means to balance the numerical inferiority but also destroy the morale of the opponent's forces. The important characteristic of a surprise attack is that it allows the initiator the choice of time, place, and method of engagement, while denying all these to the opponent. As it may make possible a quick victory at relatively low costs, surprise attack is the most important means through which a weaker initiator can expect to defeat (or avoid defeat in an encounter with) an adversary who holds more power resources.

The weaker side is conscious that time would not be on its side in an asymmetric conflict and that once the stronger side had been able to bring its full strength to bear, it would most likely lose a long drawn out war of attrition. On the other hand, Paul (ibid.: 35) concludes that 'if a weaker power expects to fight an offensive, attrition-style warfare, it may desist from engaging in war initiation'. Japanese leaders in 1941, for example, basing their calculations on the Russo-Japanese war, believed that they could fight a limited war against the United States which would be quickly terminated with a negotiated settlement that would secure Japan's territorial gains in Southeast Asia. They did not expect that the United States, after the shock of the initial Japanese victory and the prospect of waging a two-ocean war singlehandedly if Britain succumbed to Germany, would want to wage a long-term war of attrition across the Pacific (ibid.: 68).

A number of studies of the PRC's patterns of warfighting over the past 50 years suggest that China has on several occasions displayed patterns of strategic behaviour which reflect the profile of weaker state asymmetric war initiators outlined by T. V. Paul. China's decision to intervene militarily in the second stage of the Korean War in November 1950, and the conflict with the Soviet Union at Damansky-Zhenbao Island on the Ussuri River initiated by China in March 1969 stand out as clear examples of war initiation by a weaker state. In both these cases, as well as in the Taiwan Strait crises of 1955 and 1958, Beijing risked confrontation with stronger nuclear powers which could have easily escalated beyond its control.

The study of the historical record of the PRC since 1949 and published doctrinal and policy statements of Chinese officials and commentators, by Mark Burles and Abram Shulsky, suggests a somewhat different conclusion to that of either Johnston or Swaine and Tellis about the likelihood of China using military force when it is not in a position of relative material superiority over its adversaries. The principal conclusion of the Burles and Shulsky (2000: 1) study is that

> the Chinese appear to believe that they possess tactics and methods that make it feasible for them to use force even when the overall military balance is very unfavourable to them, i.e., in situations in which their use of force might otherwise have been thought very unlikely.

Ryan et al. (2003: 18) reach a similar conclusion in their survey of the PLA's warfighting experience since 1949. They consider that one could argue that 'regardless of the relative state of PLA modernization and capability vis-à-vis potential enemies ... decision makers in Beijing have themselves not been deterred from going to war when they have decided their interests were at risk'.

Burles and Shulsky (2000: 5) find that a key characteristic of Chinese use of force in actual conflict – and not just in situations where China's military power was inferior to that of its opponents – has been the importance of the element of surprise. In November 1950, for example, despite Chinese warnings against crossing the 38th parallel and although the United States was aware of the presence of Chinese troops in Korea, the Chinese attack against UN forces achieved effective tactical surprise. China's routing of Indian forces during the Sino-Indian border war of October and November 1962 was also due in large part to the Chinese forces' success in achieving tactical surprise. In both cases, it seems that the strategic effect sought by the Chinese was not only military advantage but also psychological or political shock (ibid.: 10). In the Korean case, Beijing hoped to inflict such a large and demoralising defeat on UN forces that the resulting political shock would lead the United States to withdraw from the Korean Peninsula altogether. In the Indian case, the aim was to force New Delhi to abandon its 'forward policy' and, in effect, the territorial claims it was meant to support (ibid.: 11). The strategic objective of China's invasion of Vietnam in 1979 – to 'teach Vietnam a lesson' – was also to administer a political shock which would force Hanoi to withdraw its forces from Cambodia and rethink its policy of alliance with the Soviet Union.

A second characteristic of the PRC's use of force, particularly in relation to Beijing's pursuit of its territorial claims, has been a clear pattern of opportunistic timing. A key factor, according to Burles and Shulsky (ibid.: 15), has been the isolation of China's target state from allies and other potential sources of support. In 1974 China seized the Crescent Group in the Paracel (Xisha) Islands from South Vietnam at a time when Saigon had been effectively abandoned by Washington and when Hanoi was in no position to complain. In 1988 China seized Johnson Reef in the Spratly (Nansha) Islands from Vietnam at a time when Soviet support for Vietnam was faltering and when Moscow was hoping for a rapprochement with Beijing. In 1995, China occupied Mischief Reef at a time when the Philippines were relatively isolated from the United States following the cancellation of the leases on US military bases at Clark Field and Subic Bay. A third characteristic in the PRC's use of force has been to precipitate a crisis in order to make gains that might not have otherwise been achievable, to probe the reactions of adversaries or to deter them from pursuing policies inimical to China's interests (ibid.: 16). Thus, one explanation for the 1958 Taiwan Strait crisis is that it served to rally the Chinese people behind the regime's launch of the Great Leap Forward. Beijing precipitated the 1954–5 Taiwan Strait crisis in order to break up an incipient US–Taiwanese alliance and pre-empt the conclusion of a mutual security treaty between Washington and Taipei. China's ambush of Soviet border troops on Damansky-Zhenbao Island in March 1969 was likely to have been an attempt to deter Moscow from increasing military pressure on the border by demonstrating that Beijing was willing to risk war rather than make the political concessions that Moscow was seeking (ibid.: 21).

China's intervention in the Korean War, its initiation of the Taiwan Strait crises, its invasion of Vietnam and the Damansky-Zhenbao Island incident are evidence of Beijing's willingness to risk the use of force against a stronger power or the client of a stronger power. In common with the cases of asymmetric warfare initiation examined by T. V. Paul, Burles and Shulsky (2000: 41) suggest that the strategic effects that Beijing sought to achieve by the use of force on these occasions included:

- use of surprise to create psychological shock;
- inflicting casualties to create political pressure on the opponent;
- creation of tension to divide the opposing alliance or to create political problems for the opponent;
- creation of a fait accompli, presenting the opponent with a choice between acquiescence and escalation.

Burles and Shulsky conclude that China's military inferiority would not necessarily deter Beijing from using force against a superior adversary in the future if Chinese decision-makers judged that such action stood a good chance of achieving the political effects they were seeking. Moreover, consistent with the predilection towards the pre-emptive use of force and the use of surprise tactics which is evident from the analysis both of Chinese strategic culture and historic practice, and in conformity with the patterns of strategic behaviour of asymmetric war initiators described by T. V. Paul, they suggest (ibid.: 74) that:

> In any conflict or potential conflict with the United States, China, understanding that it is the generally weaker party, would have to look for asymmetric strategies that would provide leverage against the United States ... It would seek ways to exploit US vulnerabilities and to prevent the United States from bringing its superior forces to bear. Fundamentally, China would seek to create a *fait accompli* thereby forcing the United States, if it wished to reinstate the *status quo ante*, to escalate the level of tension and violence.

Evidence for this kind of thinking in the PLA comes from David Shambaugh's (2002: 310) May 2000 interview with a PLA senior colonel in the Institute of Strategic Studies at the National Defence University in Beijing who, in response to a question about the PLA's capability to seize Taiwan, observed:

> We are accustomed to asymmetric war – we may not possess superiority in weapons over Taiwan, but our whole history of the PLA is to achieve victory over superior forces. The gap today is not nearly as great as in the Korean War. The PLA is not well prepared for war

against Taiwan, but we have never been well prepared for past wars and have always met our objectives. Our capabilities for information warfare and electronic warfare are not strong, but more likely are missile and air attacks and possibly blockade.

In addition to the more general factors inherent in the distribution of power in the international political structure, Paul identified four proximate or more immediate causal factors or conditions in the initiation of wars by weaker states. These four conditions are (1) that the key decision-makers in the weaker state believe that they can achieve their political and military objectives through the employment of limited aims/fait accompli strategies; (2) the formal or tacit support of a great-power ally or the absence of alliance support for the stronger opponent; (3) the possession of short-term offensive capabilities which offer the promise of success from a first strike; (4) changes and instability in the domestic power structure of the weaker state which increase the influence of militaristic groups in the decision-making process. The domestic factor plays a more significant role in war initiation if the regime's legitimacy or popularity are low.

As we have seen, at least two of these causal conditions for war initiation are likely to be present in a future Taiwan Strait crisis: the evidence from the examination of Chinese strategic culture and the strategic behaviour of the PRC indicates a predisposition towards limited aims/fait accompli strategies on the part of its leaders; and the possession of a strong and effective tactical submarine capability – the offensive weapon of choice for a weaker state in a maritime conflict. The two remaining conditions – instability in the domestic power structure and the ascendancy of groups inclined to militaristic policies in national decision-making and the alliance relations of the two belligerents – certainly have the potential to play a role in a future Taiwan Strait crisis. The 1995–6 crisis showed how the leadership in Beijing, its authority and legitimacy under pressure from the need to respond robustly to President Lee Teng-hui's statements and his visit to the United States, chose to use force to impose a return to the status quo ante. During periods of social unrest and party weakness in China, the military have tended to play a greater role in high-level party affairs (Shambaugh 2002: 18): 1996 was such a period. Even though the PLA was remarkably restrained in its reactions in the lead-up to and aftermath of the Taiwanese presidential elections in March 2004, there is every possibility that PRC decision-makers could come under similar pressure again – perhaps around 2006 as President Chen Shui-bian moves to fulfil his promise to reform Taiwan's constitution – at a time when the CCP's authority and legitimacy are already under great pressure as a result of economic difficulties and consequent social discontent.

In the meantime, in the wake of its military intervention in Iraq, the United States is experiencing a crisis of legitimacy of its global leadership

(Kagan 2004), which has led to an unprecedented strains on US security alliances even among its traditionally strongest allies. As John Ikenberry (2004: 7) has put it,

> it is hard to think of another instance in American diplomatic history where a strategic wrong turn has done so much damage to the country's international position – its prestige, credibility, security partnerships and goodwill of other countries – in such a short time, with so little to show for it.

At the same time, never before in the history of the PRC has Beijing enjoyed more cordial relations with its regional neighbours, a consequence in large part of its diplomatic offensive to offset Washington's dominant influence in Asia. The November 2002 agreement on proposals to create a free trade area between China and the Association of Southeast Asian Nations (ASEAN), and the code of conduct on handling territorial disputes in the South China Sea represent the fruits of these efforts. Beijing's June 2003 proposal for a regional 'Security Policy Conference' under the auspices of the ASEAN Regional Forum is further evidence of the PRC's efforts to improve its international standing and create a regional political environment that would be, if not supportive, at least neutral in the event of a confrontation with Taiwan and the United States. In June 2004, for example, Dr Eric Teo (2004), Deputy Head of the Singapore Institute of International Affairs, warned Taipei explicitly not to expect ASEAN support in a conflict with the PRC.

Paul (1994: 23) observes that the sheer presence of these factors is not sufficient to explain war initiation in all historical contexts, but notes that 'their arrival in different configurations in asymmetric conflict situations may increase the probability of war initiation by the weaker belligerent state'. Moreover, 'the more these conditions are present, the more likely that war would occur in an asymmetric conflict relationship'. Nevertheless, the PRC's possession of a powerful and effective undersea warfare capability, together with its leadership's likely propensity for limited aims/fait accompli strategy and the PLA Navy's offensive operational doctrine should be considered as factors favouring a surprise pre-emptive attack against units of a US carrier battle group by the PLA Navy's tactical submarines in a future Taiwan Strait crisis.

As Paul Bracken (1999: 129) has remarked, the greatest military surprises in American history have all been in Asia, and all from a tradition of surprise and stealth: Pearl Harbor, the Chinese intervention in Korea and the Tet Offensive in Vietnam. Each of these surprises, moreover, were classic cases of asymmetric war initiation by weaker opponents of the United States. A more apt historical analogy to a possible future Chinese submarine attack against major surface units of the US Navy might be that of the sinking of the US battleship *Maine* on 15 February 1898 in Havana harbour, supposed at the time to have been caused by a submarine mine, an incident which touched off the Spanish–American War and eventually led to the United States supplanting Spain as the colonial ruler of the Philippines and to the establishment of the United States as a permanent military presence in the western Pacific.

12

CONCLUSION

In 1993, Russia had inherited around 205 tactical nuclear and diesel-electric submarines in addition to its 55 strategic ballistic missile submarines (SSBN) from the former Soviet Union (Van der Vat 1995: 347).[1] By 2003, this number had been reduced to around 35 tactical and 13 strategic submarines (IISS 2003: 90). Since the end of the Cold War therefore, with its 67 boats, the PRC has become the possessor of the world's largest operational tactical submarine fleet. A large number of these boats are old, noisy and obsolete. Even so, with the handful of modern, quiet, well-armed *Kilo*- and *Song*-class boats, its submarine fleet is the PLA Navy's most formidable force (Cole 2003a: 135). Taken as a whole, the PLA Navy may be no match for the next largest navy in the western Pacific, that of Japan's self-defence forces, let alone the United States Navy, the Pacific's most powerful naval force. The importance of its submarine arm in its overall force structure marks the PLA Navy as primarily an instrument of strategic defence. But at the tactical and operational level, its undersea warfare capability makes the PLA Navy a force to be reckoned with – even for the US Navy. This is especially true in the narrow seas of the East Asian littoral where China's naval forces could exploit the natural advantages of a coastal navy over the United States' blue-water fleet.

Yet, for all that the PLA Navy's increasingly capable submarine arm poses a serious challenge to the US command of the seas of East Asia, and that Chinese submarines have the potential to transmute tactical success into lasting political and strategic advantage for Beijing, it must be remembered that submarines, by their nature, are able only to deny the free use of the sea to others, not command it for their own state's use. For that, there is no substitute for surface combatants. This was a point made by Bernard Brodie in 1944 (167–8) when, echoing Castex's opinion, he passed judgement on the strategic value of the submarine in the aftermath of the great submarine and anti-submarine campaigns of the Second World War:

> the submarine has not in two great trials proved the *decisive* weapon on the seas. It has lessened somewhat but by no means demolished

the significance of surface superiority. It has remained from first to last a raider, dealt with by methods comparable to those which have always been used against raiders, and not a means of wresting command of the seas. Its utility is enormous. In our own hands it was no doubt destined to hasten considerably our victory over Japan; but even against that industrially weak nation naval command cannot be won by the submarine alone.

Yet in 1958, the year after the launch of *Sputnik*, as the outlook appeared increasingly grim for the United States' ability to prevail in the contest for strategic supremacy with the USSR, Brodie's (1958: 232) assessment of the threat posed by the Soviet Navy's submarine fleet was not so sanguine:

> The Soviet Navy today has respectable surface power with a fine fleet of modern cruisers and destroyers, but in view of world geography and other considerations, this factor in itself could hardly be of importance to us. What presents the real naval menace of the Soviet Union is its great and growing submarine fleet, generally estimated at this writing to involve well over 300 of the larger sea-going units, as well as lesser craft.

Today, it is the PLA Navy's submarine fleet rather than that of the Soviet Navy which presents the greatest challenge to US command of the seas in the region which has become 'the new strategic centre of gravity in international politics' (Tellis *et al.* 1998: 43). The PLA Navy undoubtedly possesses some impressive surface capabilities, notably its new *Sovremenny* destroyers with their advanced anti-ship cruise missiles. But it is a long way from being able to pose a credible challenge to the US Navy's overall maritime supremacy in Asia. At the same time, if the PRC's large and increasingly capable tactical submarine fleet does not yet have the ability to prevent the US Navy from using the seas surrounding Taiwan with impunity, it is rapidly acquiring that ability. In a future Taiwan Strait crisis this ability could buy Beijing precious time to achieve its strategic aims vis-à-vis Taiwan. And even if a decisive Chinese naval victory is hardly likely, should the PLA Navy's submarines be lucky enough to cause serious damage to valuable ships of the US Pacific Fleet, the prestige of American arms would suffer a serious blow. So, while the PLA Navy's submarine fleet may be an instrument of strategic defence, tactically and operationally the potential threat that Chinese submarines pose to the ability of the United States Navy freely to exercise command of the littoral waters of East Asia does carry risks for the United States' strategic position in the Asia-Pacific region.

In seeking the means to deny the US Navy its freedom of action in Chinese littoral waters, China has conformed to a pattern that established itself from the moment the submarine first proved itself an effective instrument for inferior navies to attain a limited, transient measure of sea control when they were otherwise too weak to wrest the command of the surface of the sea from the dominant sea power. The submarine is, as Doenitz put it, 'ideal as a tactical weapon of offence' for navies whose inferiority compels them on to the strategic defensive at sea and therefore obliges them to adopt, in Castex's words, a strategy of 'offensive defence' if they wish to deter, preempt or constrain an attack by the superior sea power. In the past, these inferior navies have invariably belonged to land powers. For these weaker continental navies, the submarine holds the promise of some measure of tactical, operational or even strategic success in spite of the command of the sea enjoyed by the navy of the superior sea power. Moreover, the historical experience suggests that the hope of success invested by weaker navies in their submarine forces may not be misplaced. As Owen Cote (2003: 89) has observed:

> In both world wars, submarines ... tended to win the first battles between pro and anti-submarine forces. ASW success for the major naval powers depended upon urgent wartime adaptation at the technical, operational, and tactical levels. Even for those powers that were relatively quick to adapt, as were the British in World War I and both the British and the United States in World War II, their eventual ASW success was tempered by the radically asymmetric levels of effort by the contestants in favor of the submarine. The one major naval power that failed to adapt to a wartime submarine challenge, the Imperial Japanese Navy in World War II, lost control of its oceanic supply lines to American submarines with catastrophic consequences.

This is a lesson that PLA strategists seem to have taken to heart. From its foundation in 1950, the PLA Navy's force structure has been weighted heavily in favour of its submarine arm. Given the similarities of their political, economic and strategic circumstances, it made sense at that time for the newly founded People's Republic's reconstituted navy to follow the Soviet model of maritime strategy and naval force structure. Soviet strategic planners also saw great value in submarines as an effective means for a weak and impoverished land power to deter or defend against attack by the stronger naval forces of an opposing sea power. This rationale for maintaining a strong submarine arm remains just as valid for the PLA Navy today as it was in 1950, although the primary threat oscillated from the United States to the Soviet Union and back to the United States again during the half-century of its existence – with Chinese strategic planners culturally conditioned always to keep a weather eye out for a resurgence of Japanese offensive naval power.

The strategic and operational logic which governs the choice of submarines as a primary component of the PLA Navy's force structure bears a certain similarity to that which emerged from the Imperial German Navy's experiences with submarines in the confined and shallow waters of the North Sea during the First World War. In strategic and geographic circumstances not unlike those the PLA Navy faces today in the confined and shallow waters off the China coast and around Taiwan, the German Navy sought by various operational stratagems to ambush the superior forces of the British Grand Fleet, in the hope of achieving sufficient tactical successes against the Royal Navy's capital ships to reduce the odds of achieving a strategic victory in a full-scale fleet engagement. In the event of a war precipitated by a political crisis over the status of Taiwan, the PLA Navy's operational objective would most likely be to use its most capable submarines to try to sink or seriously damage US aircraft carriers or other valuable assets. The *Kilo*-class submarines' wake-homing torpedoes offer the PLA Navy its best chance of success in achieving this objective. Modern stand-off, fire-and-forget weapons mean that even less experienced and competent submarine crews can achieve success against better trained, equipped and experienced navies (Cote 2003: 79). Gray (1992b: 76) was probably thinking of Imperial Germany or the Soviet Union rather than China when he described the logic of this approach, although it applies equally well to the PRC in the first decade of the twenty-first century:

> With its second-class naval power more or less well protected in coastal bastions, a land power can entertain the hope that the sea power enemy will accept imprudent risks in order to attempt to bring on a decisive fight at sea. The probable fact that the navy of the sea power enemy will be heir to an offensive tradition in strategy should be key to the prospects for success of this idea. If the commanding power at sea declines to hazard its battle fleet in circumstances of great tactical disadvantage, still that power may have its superiority whittled away by the attrition suffered through overconfidence in a series of relatively minor engagements.

The ability of the PLA Navy's submarine force to pose a credible threat to the US Navy's most valuable assets is, in itself, a powerful deterrent to US naval operations in the seas surrounding Taiwan. The very existence of this force may be sufficient to restrict the freedom of action of US carrier battle groups, and to reduce their ability to influence events in the Taiwan Strait by obliging them to maintain their distance from the likely operating areas of the PLA Navy's submarines. In 1914, it was the very existence of the Imperial German Navy's U-boats that caused the Royal Navy to abandon its long-held strategy of close blockade of German North Sea ports in favour of a more distant blockade of the northern exit of the North Sea from the

Grand Fleet's new base in the Orkney Islands. Nearly a century later, the very existence of the PLA Navy's submarines in the seas around Taiwan would, at the very least, impose significant costs on the US Navy in terms of the resources it would have to devote to ASW.

More broadly, the emerging contest for control of the seas around Taiwan, in which the primary protagonists are the PLA Navy's tactical submarine fleet and the US Navy's carrier battle groups, has much in common with the contest in the 1950s and 1960s in which Soviet attack submarines directly challenged US carrier strike groups' command of the sea. Both contests seem to provide evidence to support John Keegan's (1993: 266) contention that 'by the end of the Second World War, indeed well before its end, the submarine and the aircraft carrier had established themselves indisputably as the dominant weapons of war at sea'. The last large United States aircraft carrier to be destroyed in battle at sea was the USS *Hornet* at Santa Cruz in October 1942. If in the course of a conflict in the Taiwan Strait PLA Navy tactical submarines were to succeed, admittedly against the odds, in achieving their aim of destroying a US Navy aircraft carrier – with heavy casualties among the 6,000 personnel aboard – it would be not only a serious military and strategic setback for the United States and its pre-eminent position in Asia, but would be a psychological and political blow to Americans of the same order as the 11 September 2001 attack on the World Trade Center. It would also incidentally strengthen the argument of those who, like Keegan (ibid.: 272), contend that 'command of the sea in the future unquestionably lies beneath rather than upon the surface'. This was certainly the direction that the virtual contest for command of the sea took during the Cold War, when strike carriers were superseded by SSBN/SLBMs as the United States' primary offensive strategic weapons systems at sea, and SSNs became the primary Soviet means of defence. The deployment by China of its new Type-093 SSN around the end of this decade and, around 2007, by the United States of the first of its four *Ohio* SSBNs converted into SSGNs armed with Tomahawk land-attack cruise missiles, could mark a similar transition from the surface to the subsurface in the Sino-US contest for control of the East Asian seas. These developments would in any case provide evidence for the argument that the future of sea power lies as much beneath the sea and in space as on its surface or in the skies above.

Consistent with the classic division in naval strategic theory between the *guerre d'escadres* and the *guerre de course* (Coutau-Bégarie 2000: 584), Castex (1997: vol. I, 278) noted that submarines have two essential roles: to attack the enemy's organised forces – its battle fleet; and to attack the enemy's lines of communications – its merchant and military transport and logistics fleets. In defining the primary missions for its submarine forces, China's strategic planners appear to have adhered to this classic dichotomy. In addition to their objective of attacking elements of the opposing fleets, the PLA Navy's submarines also have the principal role in imposing a maritime blockade on Taiwan. China's choice of its submarine arm as the primary instrument to carry out this vital strategic mission is consistent with historic patterns of submarine employment established by Germany in the First and Second World Wars, the United States in its war against Japan, and the Soviet Union during the Cold War. In each case, the submarine was regarded as the weapon of choice for naval powers which were able to exercise only limited or intermittent control of the surface of the sea and therefore unable to bring force to bear directly against the enemy's territory. In each of these cases, the submarine campaign against enemy merchant shipping was supplemented by an air-power campaign directed against enemy economic and military targets. Both campaigns were part of what Admiral Wylie (1967: 23) called a 'cumulative strategy' designed to have both an economic and a psychological effect on the enemy. The PRC has also adhered to this pattern, although, not having sufficient confidence in the ability of the PLA Air Force to establish air superiority over the Taiwan Strait, it has substituted missile power for air power as the complementary instrument of this cumulative strategy. The latest Chinese Defence White Paper, published in December 2004, gives priority to the navy, the air force and the Second Artillery Force in developing the PLA's operational strength.

China's conformity with past patterns of strategic behaviour is evidence of the operation of a universal and timeless logic guiding the policy process of matching strategic ends and means which results in its submarine arm playing such a prominent role in the PLA Navy's force structure. But this logic of consequences is supplemented by a logic of appropriateness – typical of decision-making organisations – which complements and reinforces the strategic choices reached by the application of the former logic. This logic of appropriateness injects the influence of their strategic culture into the decision-making process of China's military planners concerned with naval force structure and operational strategy. Strategic culture helps to guide and shape decisions relating to the acquisition and employment of the PLA Navy's submarine arm. The analysis of Chinese strategic culture also throws some light on the patterns of thought of the Chinese military mind in a way that helps to understand some of the possible actions that could grow out of those thought patterns in the context of an actual war.

China's classical strategic culture, either directly or transmitted and modernised through the medium of Mao Zedong's strategic precepts, is another primary influence on the way in which PLA strategic planners and policy-makers think about submarine warfare. Although these ancient traditions spring from an almost exclusively terrestrial experience of warfare, they nevertheless contain much that is relevant to littoral and undersea warfare where, unlike warfare on the surface and the open ocean, 'terrain' has significant tactical importance. The history of the use of submarines shows that their essential comparative advantage over surface combatants is their stealth, and this inherent quality lends itself to their use as offensive weapons for surprise, and often pre-emptive attacks by weaker forces against stronger adversaries. The inherent qualities of the submarine make it a particularly attractive weapon for Chinese strategists whose strategic culture emphasises offensive operations and tactics within a generally defensive strategy, and which favours pre-emptive attacks taking advantage of the elements of surprise and manoeuvre and making strategic use of terrain. In every significant maritime conflict since the submarine came of age as a formidable weapon of naval warfare in 1914, it has proved itself to be the asymmetric weapon of choice for inferior navies. Hopelessly outclassed by the US Navy in just about every other dimension of naval warfare, a large and capable submarine arm at least gives the PLA Navy reasonable grounds for hoping to score sufficient tactical success to achieve a strategic victory. Even if they fall short of this goal, Chinese strategists may well reason as Castex (1997: vol. I, 319) believed that the Germans might have done after their valiant but ultimately vain submarine campaign during the First World War:

> The commander of the surface will eventually prevail. We know that better than anyone. But even so, through continual activity, we will have saved morale and honour in the present and won for the future the respect of other peoples and a prestige whose political repercussions could be significant. This is all that is required for the submarine to be of immense interest for navies whose means condemn them to not having in every case that superiority or equality on the surface which is the attribute of only the few rare great navies of the world.

The picture that emerges from this examination of the reasons behind Beijing's determination to maintain and develop a powerful and effective submarine force is consistent with the conclusions of other analysts that the Chinese political and military leadership is pursuing a policy of strategic modernisation of the PRC's armed forces. Instead of attempting systematically to modernise the complete spectrum of the PLA's capabilities, Beijing is, as Mark Stokes (1999: 5) puts it:

> concentrating on the development of doctrine and systems designed to enable targeting of adversarial strategic and operational centers of gravity, and defend its own, in order to pursue limited political objectives with an asymmetrical economy of force … China's strategic modernization, if successful, will enable the PLA to conduct operations intended to directly achieve strategic effects by striking the enemy's centers of gravity. These operations are meant to achieve their objectives without having to necessarily engage the adversary's fielded military forces in extended operations. Strategic

attack objectives often include producing effects to demoralize the enemy's leadership, military forces, and population, thus affecting an adversary's capability to continue the conflict.

The acquisition and development of military capabilities which would enable the pursuit of limited political objectives with an asymmetrical economy of force is consistent with the pattern of strategic behaviour of weak states, including China, which have in the past initiated conflicts with stronger adversaries. A key factor in weaker state war initiation is the belief of its political and military leadership in the efficacy of limited aims/fait accompli strategies to win their strategic objectives. The strategic behaviour of the PRC over the past half-century shows that limited aims/fait accompli strategies have been an enduring feature of Beijing's use of force in the conduct of its international relations. The possession of an offensive capability that holds the promise of successful surprise attacks against the military forces of a superior adversary is another key causal factor in the initiation of asymmetric conflicts by weaker states. The PRC's possession of advanced undersea warfare capabilities could therefore be a destabilising factor in a future Taiwan Strait crisis and could precipitate a conflict between the PLA and US military forces – a conflict which, unlike any of the conflicts in which US forces have engaged since the end of the Cold War, would be first and foremost a maritime conflict between the naval forces of the two sides. For, to repeat Thomas Schelling's (1966: 234) observation, 'weaponry does affect the outlook for war and peace ... [It] can determine the calculations, the expectations, the decisions, the character of a crisis, the evaluation of danger and the very processes by which war gets under way.'

Ever since sea power became a major factor in the international politics of East Asia, the island of Taiwan has had a vital strategic significance. Since the Treaty of Shimonoseki gave control of the island to Japan following its victory over the Peiyang Fleet in 1894, the island has been controlled by the dominant maritime power of the day, or its ally. If, as classical sea power theorists would contend, the priority accorded by Beijing to the development of its naval forces is evidence of the re-emergence at the beginning of the twenty-first century of the historical pattern of strategic competition for primacy between a dominant continental power and the pre-eminent sea power, then the control of the island of Taiwan will acquire an even greater strategic significance for both parties in the future than it already has in the past. In the nuclear age, as the Cold War confrontation so clearly demonstrated, nuclear forces ultimately underwrite great-power strategic competition. Thus if Beijing is to challenge Washington successfully for primacy in East Asia, China will have to acquire a credible second-strike nuclear deterrent capability. Its current force of silo-based, liquid-fuelled ICBMs and even its next generation of road-mobile, solid-fuelled missiles are likely to become too vulnerable to provide the PRC leadership with sufficient confidence in their ability to deter the United States from launching a pre-emptive first strike. A submarine-launched ballistic missile force capable of striking targets in the United States homeland is Beijing's best option for a credible Chinese nuclear deterrent. However, unless China is able to deploy an SLBM with a range far greater than the 8,000 kilometres of its future JL-2 missile (expected to be deployed by the end of the current decade), its ability to pose a credible nuclear threat to the United States will depend on the PLA Navy's ability to assert sufficient control of the East Asian seas at the very least beyond the first island chain in order to secure the passage of its SSBNs to their mid-ocean launch zones. But as long as the island of Taiwan is in hostile hands, the PLA Navy cannot be assured of unrestricted access to the open waters of the Pacific Ocean for its SSBN force, and China's ability to pose a credible nuclear threat to the United States must be in doubt. Thus the future of Taiwan is not only a political issue of major significance to both Beijing and Washington, but it is also at the heart of the great-power competition for strategic supremacy in East Asia.

If the opinion that Washington would be unwilling to trade Taipei for Los Angeles, embodied in General Xiong's implied nuclear threat, is indicative of a general belief among China's military and political leaders that the strategic stake in the future of Taiwan is greater for Beijing than it is for Washington, then there is a risk of a serious miscalculation on the part of the Chinese leadership. Such a belief, if genuinely held, would betray a failure on the part of China's leaders to grasp that basic tenet of statecraft embodied in E. H. Carr's (1946: 110) moral 'that foreign policy never can, or never should, be divorced from strategy' – a moral also epitomised in Sun Zi's maxim that 'the conduct of war is vital to the nation' (Chow-Hou 2003: 7). This is because a state's foreign policy is ultimately concerned with its military strength or, more accurately, with the ratio of its military strength to that of other countries (Carr 1946: 110). In the context of the international politics and strategic geography of East Asia, the state – or alliance of states – which controls Taiwan thereby increases its military strength immensely. For Taiwan, as a key component of the island chain on the western rim of the Pacific Ocean, is vital to the projection of sea power westwards to the continent of East Asia, or eastwards to the continent of North America. Seen from this perspective, the security of Los Angeles is indeed at stake in the control of Taiwan, and thus it has to be counted among the vital strategic interests of the United States. The leadership in Beijing would therefore be in serious error if they believed that Washington would not run the risk of war to protect its strategic stake in Taiwan. For, as E. H. Carr (1946: 111) pointed out:

Few important wars of the last hundred years seem to have been waged for the deliberate and conscious purpose of increasing either trade or territory. The most serious wars are fought in order to make one's own country militarily stronger, or, more often, to prevent another country from becoming militarily stronger.

That said, the last thing that either China or the United States would want is to go to war with one another over Taiwan. There is no reason to doubt the sincerity of China's leaders when they maintain that the future of their nation depends critically on the avoidance of a war which could reverse much of the economic and social progress made by their nation over the course of the last quarter of a century. But there is nowhere else in the world today where there is such a great potential for a war between great powers as in the seas surrounding Taiwan – a potential highlighted by the intrusion of a PLA Navy *Han*-class submarine into Japanese territorial waters in November 2004.

A. J. P. Taylor's (1971: ix) assertion that the relations of the great powers have determined the history of Europe applies with equal force to the future of Asia. Whether, as John Mearsheimer (2001: xii) suggests, China and the United States are fated to clash as all great powers are in their competition for power and security, there is no doubt that the complex drama in the Taiwan Strait has many of the ingredients of a tragedy. History has many examples of wars where the principal players in the drama have been driven to war or the brink of war against their wishes and better judgement because of the actions of minor players. Neither Moscow nor Washington sought a conflict in 1973, but the actions of Egypt brought about a crisis in superpower relations more serious than any since the 1962 Cuban missile crisis. And history also has many examples of wars that come about not as a result of deliberate policy but because of misperception, miscalculation and accident.

The re-election on 20 March 2004 of President Chen Shui-bian, leader of Taiwan's Democratic Progressive Party, for a second term does not improve the prospects for an easing of the tension and the risks of war across the Taiwan Strait. His re-election confirms the entrenchment of support for independence among the Taiwanese electorate as well as the divisions within Taiwanese society between those who favour independence and those who favour an accommodation with the mainland. The net result of the 2004 election has been to exacerbate instability both within Taiwan and across the Taiwan Strait.

At the same time, Beijing is continuing to develop the means to effect a military solution to the Taiwan issue. In March 2004, the PRC's Finance Minister announced a defence budget for 2004 of US$25 billion, an 11.6 per cent increase over the previous year (Cody 2004). These figures do not include funds allocated for acquisitions or research and development. The Pentagon estimates that the Chinese annual defence budget is actually between $50 and $70 billion, putting the PRC in third place behind the

United States and Russia as the world's largest spenders on defence (US Senate 2004). During the 12 months preceding the announcement of the 2004 defence budget, the PRC had also begun the construction of more than 70 new vessels for the PLA Navy (Cody 2004).

Meanwhile, in Europe, France and Germany are trying to convince their European Union partners to lift the embargo on the sale of arms to the PRC imposed after the Tiananmen events of 1989. If they succeed, the way will be open to Beijing to acquire sophisticated military equipment such as radars, sonars, torpedoes and naval anti-air defence systems. This is likely to accelerate and intensify the arms race across the Taiwan Strait, which can only serve to aggravate the tension between Beijing and Taipei, with negative consequences for the security and stability of the whole Asia-Pacific region. There is therefore a strong likelihood that a second presidential mandate for Chen Shui-bian will usher in a period of increased tension and heightened risk of war in the Taiwan Strait.

NOTES

INTRODUCTION

1 ' Srikanth Kondapalli (2001: xv) comments pertinently on this somewhat puzzling lacuna in PLA studies:

> For any country, its relatively developed naval forces constitute its power projection elements abroad. By this logic, the PLAN should have attracted wider attention among China-watchers than it has. Of nearly forty specialists in the world reflecting on various aspects of the Chinese military today, only a few have thought it worthwhile to focus on China's naval forces.

1 CHINA'S TACTICAL SUBMARINE FLEET

1 According to A. D. Baker III (2004: 37), the number of tactical submarines in the PLA Navy's fleet is somewhat different: 32 *Romeos*; 3 nuclear *Han*-class boats; 5 *Songs*; 4 *Kilos* and 19 *Mings*, for a total of 63 tactical boats.

2 THE GEOPOLITICAL CONTEXT

1 See, for example, Nathan and Ross (1997); Mearsheimer (2001); Swaine and Tellis (2000); Roy (2003); Brzezinski (2004).
2 Admiral Raoul Castex (1997) *Théories stratégiques*, Paris: Editions Economica. Only selections of Castex's massive seven-volume work, originally published between 1929 and 1935, have been translated into English (R. Castex [1994] *Strategic Theories*, trans. and ed. E. Kiesling, Annapolis: Naval Institute Press). Quotations from *Théories stratégiques*, and all other quotations from works in French in the present work, are the author's own translations.
3 The design information for the W-88 thermonuclear warhead used on the US Trident D5 SLBM allegedly acquired by China could provide significant assistance to Chinese weapon designers working towards the development of an intercontinental-range SLBM (United States Senate 1999).

3 CHINA'S NEW MARITIME STRATEGY

1 The expressions 'brown water', 'green water' and 'blue water' to classify different types of navies and their capabilities, missions and operational environments are

often used somewhat loosely. Michael Lindberg and Daniel Todd (2002: 196) offer useful clarification of these concepts:

> geographical criteria can be divided into two categories, namely, operational environment and what is commonly referred to as 'reach', the distance from home that a navy can effectively operate. Operational environment can be broken down into the broad categories of 'blue water' and 'non-blue water'. The latter category can be broken down further into 'green water' and 'brown water', the former referring to offshore, coastal waters and the latter to the waters of inland rivers. The norm is to associate power-projection navies with blue water, coastal and territorial defense navies with either green or brown water, and constabulary navies with green water. To further clarify the geographical classification of navies, the concept of reach is used. Based upon the geographical theory of decay of distance, a loss-of-power gradient is offered to illustrate the fact that the mission capability of most navies declines as they operate at ever greater distances from their home base. Consequently, it can be rightly assumed that blue-water, power-projection navies have much greater reach and therefore possess much greater capabilities than green-water, coastal defense or constabulary navies. Thus, gradations in reach, once plotted as a negatively sloping line called the 'loss-of-power-gradient', are tantamount to divisions between types of navies. These divisions yield several additional distance-specific types of blue-water navies, including global-reach, limited global-reach, and regional power-projection navies. This division also adds to the clarity of green-water navies with regional offshore and inshore coastal defense and constabulary navies.

It is clear from this explanation that the PLA Navy of today is perhaps best described as a 'green-water' navy in the early stages of a process of gradual transformation towards becoming a regional power-projection blue-water navy.

2 As Michael A. Glosny (2004: 128) points out, however, the critical factor in assessing the likelihood of Beijing achieving its political goals by means of a naval blockade of Taiwan is the will of the people of Taiwan to resist. Glosny's careful analysis of a hypothetical PRC submarine blockade of Taiwan leads him to conclude that:

> although a PRC submarine blockade could impose costs on Taiwan, the threat of a successful blockade is overstated. Even using assumptions very favourable to the PLA Navy (PLAN), a blockade would likely inflict only limited damage on Taiwan. Unless this damage was sufficient to force Taiwan to collapse quickly, a PRC submarine blockade would most likely not be successful. Moreover, there are good reasons to suggest that when confronted by such an attack, Taiwan would stand firm in response.

5 MARITIME STRATEGIC THEORY AND THE LOGIC OF CHINA'S SUBMARINE FLEET

1 Writing in 1985, Hervé Coutau-Bégarie (19) considered that after Castex, the twentieth century had produced no true naval strategic theorists, but only historians such as Stephen Roskill, Arthur J. Marder and Paul M. Kennedy. He cites Bernard Brodie's 1965 comment on the paucity of naval strategic theorists: 'the

rare works published after Mahan, Corbett and Castex were simply updates of their works'.

9 INFLUENCE OF THE SOVIET EXPERIENCE ON THE PRC'S MARITIME STRATEGY

1 Although the proposed means were similar, the doctrine of the Soviet 'New School' differed from that of the French '*jeune école*' which flourished from 1885 to 1895 (Coutau-Bégarie 1985: 173). The latter advocated an offensive strategy, based on unrestricted attacks on British merchant ships as a means of economic warfare, while the former proposed a purely defensive strategy to counter expected enemy naval attacks in home waters (Ranft and Till 1983: 86).

12 CONCLUSION

1 Comprising 59 nuclear hunter-killer SSNs; 2 experimental SSNs; 36 nuclear SSGN cruise-missile submarines; 8 diesel-electric SSG submarines; 80 diesel-electric SS submarines; and 10 diesel-electric SS submarines for miscellaneous use (Van der Vat 1995: 347).

REFERENCES

Agence France Press (2003) 'US Arms Offer Turned Down, Report Claims', 9 May 2003. Online. Available HTTP: www.taiwansecurity.org/AFP/2003/AFP-050903.htm (accessed 5 June 2004).

Allison, G. and Zelikow, P. (2nd edn 1999) *The Essence of Decision: Explaining the Cuban Missile Crisis*, New York: Addison-Wesley Longman.

Andronov, A. (1993) 'Kosmicheskaya Sistema Radiotekhnicheskoy Razvedki VMS SShA "Uayt Klaud"', *Zarubezhnoye Voyennoye Obozreniye* [Foreign Military Review] (ISSN 0134–921X), 7: 57–60; translated by A. Thomson, Federation of American Scientists, Military Analysis Network. Online. Available HTTP: http://www.fas.org/spp/military/program/surveill/noss_andronov.htm (accessed 23 September 2004).

Austin, G. (1988) *China's Ocean Frontier: International Law, Military Force and National Development*, Sydney: Allen & Unwin.

Baker, A. D. III (2004) 'World Navies in Review', *US Naval Institute Proceedings*, 130/3/1 (213).

Baud, J. (2003) *La Guerre asymétrique ou la défaite du vainqueur*, Paris: Editions du Rocher.

Beaufre, A. (1963) *Introduction à la stratégie*, Paris: Armand Colin.

Behn, S. (2004) 'US to Build 8 Subs in Deal with Taiwan', *Washington Times*, 29 September 2004. Online. Available HTTP: http://www.washingtontimes.com/world/20040929–123355–3804r.htm (accessed 30 September 2004).

Betts, R. K. (1993) 'Wealth, Power, and Insecurity: East Asia and the United States after the Cold War', *International Security*, 18 (3): 34–77.

Blair, C. (1975) *Silent Victory: The US Submarine War against Japan*, New York: Lippincott.

Booth, K. (1979) *Strategy and Ethnocentrism*, London: Croom Helm.

Booth, K. and Trood, R. (eds) (1999) *Strategic Cultures in the Asia-Pacific Region*, London: Macmillan.

Borik, F. C. (1996) 'Sub Tzu and the Art of Submarine Warfare', in M. A. Somerville (ed.) *Essays on Strategy*, Washington, DC: National Defense University, pp. 3–41.

Børresen, J. (1994) 'The Seapower of the Coastal State', in G. Till (ed.) *Seapower: Theory and Practice*, Ilford: Frank Cass.

Bracken, P. (1999) *Fire in the East: The Rise of Asian Military Power and the Second Nuclear Age*, New York: HarperCollins.

Brisset, J.-V. (2002) *La Chine: une puissance encerclé*, Paris: Presses Universitaires de France.

Brodie, B. (1942; 3rd edn 1944), *A Guide to Naval Strategy*, Princeton: Princeton University Press.

—— (1942; 4th edn 1958), *A Guide to Naval Strategy*, Princeton: Princeton University Press.

—— (1959) *Strategy in the Missile Age*, Princeton: Princeton University Press.

Brooke, J. (2004) 'Japanese Island Tries to Evade Flight Path', *New York Times*, 20 September 2004. Online. Available HTTP: http://www.nytimes.com/2004/09/20/international/asia/20japan.html (accessed 22 September 2004).

Brzezinski, Z. (2004) *The Choice: Global Domination or Global Leadership*, New York: Basic Books.

Bull, H. (1976) *Seapower and Political Influence*, Adelphi Paper No. 122, London: International Institute for Strategic Studies.

Burles, M. and Shulsky, A. (2000) *Patterns in China's Use of Force*, Santa Monica: RAND.

Bussert, J. C. (2003) 'Chinese Submarines Pose a Double-Edged Challenge', *Signal Magazine*. Online. Available: HTTP: http://www.afcea.org/signal/articles/anmviewer.asp?a=93&Z=22 (accessed 4 October 2004).

Butler, J. D. (2004), 'Building Submarines for Tomorrow', *US Naval Institute Proceedings*, 130/6/1 (216) 53.

Cabestan, J.-P. (2003) *Chine–Taiwan: la guerre est-elle concevable?* Paris: Economica.

Cable, J. (1995) 'Une stratégie maritime sur mesure', in H. Coutau-Bégarie (ed) *La Lutte pour l'empire de la mer*, Paris: Economica.

Carr, E. H. (1939; 2nd edn 1946) *The Twenty Years' Crisis 1919–1939: An Introduction to the Study of International Relations*, London: Macmillan.

—— (1961) *What is History?* London: Macmillan.

Castex, R (1929–35; 1997) *Théories stratégiques*, 5 volumes, Paris: Economica.

Cavas, C. P. (2004a) 'Boeing aims to Beat MMA Production Schedule', *Defense News*, 21 June.

—— (2004b) 'US to "Borrow" Swedish Sub for Training', *Defense News*, 27 September.

Chauprade, A. (2001) *Géopolitique: constantes et changements dans l'histoire*, Paris: Ellipses.

Cheng, D. (2003) 'The Chinese Space Program: A 21st Century "Fleet in Being"?', in J. Mulvenon and A. D. Yang (eds) *A Poverty of Riches: New Challenges and Opportunities in PLA Research*, Santa Monica: RAND. Online. Available: HTTP: http://www.rand.org/cgi-bin/Abstracts/ordi/getabbydoc.pl?doc=CF-189&hilite=1&qs=China (accessed 24 September 2004).

China Defence Today (2004) 'Yuan Class Diesel-Electric Submarine'. Online. Available HTTP: http://www.sinodefence.com/navy/sub/yuan.asp (accessed 4 October 2004).

Ching C. (2004) 'China Revamps Top Military Command', *Straits Times*, 10 August 2004. Online. Available HTTP: http://taiwansecurity.org/ST/2004/ST-100804.htm (accessed 12 August 2004).

Chong-Pin, L. (1988) *China's Nuclear Weapons Strategy: Tradition within Evolution*, Lexington, Massachusetts: Lexington Books.

Chow-Hou, W. (2003) *Sun Zi Art of War: An Illustrated Translation with Asian Perspectives and Insights*, Singapore: Prentice Hall.

Clark, V. (2002) 'Projecting Decisive Joint Capabilities', *Naval Institute Proceedings*, 128 (10) (1). Online. Available HTTP: http://www.usni.org/Proceedings/Articles02/PROcno10.htm#seastrike (accessed 20 March 2004).

Clausewitz, C. (1832; trans 1976; 2nd edn 1984) *On War*, edited and translated by M. Howard and P. Paret, Princeton: Princeton University Press.

CNN.com World Business Report (2003) 'China Targets National Oil Reserve', 27 November 2003. Online. Available HTTP: http://www.cnn.com/2003/BUSINESS/11/27/china.oil/index.html (accessed 2 April 2004).

Cody, E. (2004) 'With Taiwan in Mind, China Focuses Military Expansion on Navy', *Washington Post*, 20 March. Online. Available HTTP: http://www.washingtonpost.com/wp-dyn/articles/A9158-2004Mar19.html (accessed 15 September 2004).

Cole, B. D. (2000) 'China's Maritime Strategy', in S. M. Puska (ed.) *People's Liberation Army after Next*, Carlisle, Pennsylvania: Strategic Studies Institute.

—— (2001) *The Great Wall at Sea: China's Navy Enters the Twenty-First Century*, Annapolis: Naval Institute Press.

—— (2003a) 'The PLA Navy and "Active Defense"', in S. J. Flanagan and M. E. Marti (eds) *The People's Liberation Army and China in Transition*, Washington, DC: Institute for National Strategic Studies. Online. Available HTTP: http://www.ndu.edu/inss/books/Book_titles.htm (accessed 8 June 2004).

—— (2003b) 'The People's Liberation Army Navy after Half a Century: The Lessons Learned in Beijing', in L. Burkitt, A. Scobell and L. M. Wortzel (eds) *The Lessons of History: The Chinese Liberation Army at 75*, Carlisle, Pennsylvania: Strategic Studies Institute. Online. Available HTTP: http://www.carlisle.army.mil/ssi/pubs/pubResult.cfm/hurl/PubID=52/ (accessed 9 September 2004).

Corbett, J. S. (1911; reprinted 1988) *Some Principles of Maritime Strategy*, Annapolis: Naval Institute Press.

Cote, O. R., Jr (1999) 'Mobile Targets from Under the Sea: An MIT Security Studies Program Conference', December 1999. Online. Available HTTP : http://web.mit.edu/ssp/Publications/confseries/mobtarg.pdf (accessed 4 September 2004).

—— (2003) *The Third Battle: Innovation in the US Navy's Silent Cold War Struggle with Soviet Submarines*, Newport, Rhode Island: Naval War College Press.

Cote, O. R., Jr and Sapolsky, H. (2001) 'Anti-Submarine Warfare after the Cold War', Summary Report of MIT Security Studies Conference, April. Online. Available HTTP: http://web.mit.edu/ssp/ (accessed 15 September 2004).

Couper, A. (ed.) (1983) *The Times Atlas of the Oceans*, New York: Van Nostrand Reinhold.

Coutau-Bégarie, H. (1985) *La Puissance maritime: Castex et la stratégie navale*, Paris: Fayard.

—— (2000) *Traité de stratégie*, Paris: Economica.

Davis, J. K. (1998) *CVX: A Smart Carrier for the New Era*, Washington, DC: Brasseys.

Dénécé, E. (1995) 'La Liberté de navigation à travers les détroits d'Asie du sud -est', in H. Coutau-Bégarie (ed.) *La Lutte pour l'empire de la mer*, Paris: Economica.

Ding, A. S. (2003) 'The Lessons of the 1995–1996 Military Taiwan Strait Crisis: Developing a New Strategy toward the United States and Taiwan', in L. Burkitt, A. Scobell and L. M. Wortzel (eds) *The Lessons of History: The Chinese*

Liberation Army at 75, Carlisle, Pennsyslvania: Strategic Studies Institute. Online. Available HTTP: http://www.carlisle.army.mil/ssi/pubs/pubResult.cfm/hurl/PubID=52/ (accessed 9 September 2004).

Doenitz, K. (1958; English trans. 1959) *Memoirs: Ten Years and Twenty Days*, trans. by R. H. Stevens, Annapolis: Naval Institute Press.

Dreyer, J. T. (1999) 'The PLA and the Taiwan Strait', paper prepared for the International Forum of The Peace and Security of the Taiwan Strait, 26–28 July, Taipei. Online. Available HTTP: www.taiwansecurity.org/IS/Dreyer-ThePLA-and-the-Taiwan-Strait.htm (accessed 9 June 2004).

—— (2000) *The PLA and the Kosovo Conflict*, Carlisle, Pennsylvania: Strategic Studies Institute. Online: Available: HTTP: http://www.carlisle.army.mil/ssi/pubs/pubResult.cfm/hurl/PubID=70/ (accessed 16 September 2004).

Fairbank, J. K. (1974) 'Varieties of the Chinese Military Experience', in F. A. Kierman and J. K. Fairbank (eds) *Chinese Ways in Warfare*, Cambridge, Mass.: Harvard University Press.

Federation of American Scientists (FAS) (1999) 'AN/SQS-53 Sonar'. Online. Available HTTP: http://www.fas.org/man/dod-101/sys/ship/weaps/an-sqs-53.htm (accessed 14 September 2004).

——(2000a) 'Discoverer II (DII) Starlite', Space Policy Project, Military Space Programs, 24 January 2000. Online. Available HTTP: http://www.fas.org/spp/military/program/imint/starlight.htm (accessed 17 September 2004).

—— (2000b) 'MND Dismisses "Blockade" Report'. Online. Available HTTP: http://www.fas.org/news/taiwan/2000/e-05–17–00–5.htm (accessed 12 October 2004).

—— (undated) 'US Navy Ships: Submarine Warfare', Federation of American Scientists Military Analysis Network. Online. Available HTTP: http://www.fas.org/man/dod-101sys/ship.submarine.htm (accessed 17 March 2004).

Friedberg, A. L. (2000) 'Will Europe's Past be Asia's Future?' *Survival*, 42 (3): 147–60.

Friedman, G. and Friedman, M. (1996) *The Future of War: Power, Technology and American Dominance in the Twenty-First Century*, New York: St Martin's Griffin.

Friedman, N. (1995) 'Littoral Anti-Submarine Warfare: Not as Easy as it Sounds', *International Defense Review*, 28 (6): 53–7.

—— (2000) *Seapower and Space: From the Dawn of the Missile Age to Net-Centric Warfare*, London: Chatham Publishing.

—— (2001) *Seapower as Strategy: Navies and National Interests*, Annapolis: Naval Institute Press.

Gelb, L. H. (2003) 'Foreword', in *Chinese Military Power: Report of an Independent Task Force Sponsored by the Council on Foreign Relations*, New York: Council on Foreign Relations. Online. Available HTTP: http://www.cfr.org/pdf/China_TF.pdf (accessed 14 September 2004).

Gertz, B. (2001) 'China Tests Supersonic Anti-Ship Cruise Missiles', *Washington Times*, 25 September 2001. Online. Available HTTP: http://taiwansecurity.org/News/2001/WT-092501.htm (accessed 12 May 2004).

Gilbert, M. (1994) *The First World War*, London: HarperCollins.

Gill, B. and Mulvenon, J. (2000) 'The Chinese Strategic Rocket Forces: Transition to Credible Deterrence', in Conference Report on *China and Weapons of Mass*

Destruction: Implications for the United States, Washington, DC: US National Intelligence Council and Federal Research Division, Library of Congress, April.

Gilpin, R. (1981) *War and Change in World Politics*, Cambridge: Cambridge University Press.

Global Security.org (2001) 'Study: US Should Shift Military Focus in Asia Closer to Hot Spots', *Stripes and Wire Reports*, 16 May 2001. Online. Available HTTP: http://www.globalsecurity.org/wmd/library/news/china/2001/prc-010516-taiwan.htm (accessed 6 June 2004).

Glosny, M. A. (2004) 'Strangulation from the Sea? A PRC Submarine Blockade of Taiwan', *International Security*, 28 (4): 125–60.

Godwin, P. H. B. (2003a) *Change and Continuity in Chinese Military Doctrine, 1949–1999*, in M. A. Ryan, D. M. Finkelstein and M. A. McDevitt (eds) *Chinese Warfighting*, New York: M. E. Sharpe.

—— (2003b) 'China's Defense Establishment: The Hard Lessons of Incomplete Modernization', in L. Burkitt, A. Scobell and L. M. Wortzel (eds) *The Lessons of History, the Chinese Liberation Army at 75*, Carlisle, Pennsylvania: Strategic Studies Institute. Online. Available HTTP: http://www.carlisle.army.mil/ssi/pubs/pubResult.cfm/hurl/PubID=52/ (accessed 9 September 2004).

Goldstein, L. and Murray, B. (2003) 'China's Subs Lead the Way', *US Naval Institute Proceedings*, 121/3/1 (201). Online. Available HTTP: www.usni.org/Proceedings/Articles03/prgoldstein03.htm (accessed 2 February 2004).

—— (2004) 'Undersea Dragons: China's Maturing Submarine Fleet', *International Security*, 28 (4): 161–96.

Gorshkov, S. G. (1979) *The Sea Power of the State*, translated from *Morskaya moshch gosudarstva*, Annapolis: Naval Institute Press.

Goure, D. (2002) 'The Leading Edge of Transformation', *Seapower*, July. Online. Available HTTP: http://www.navyleague.org/sea_power/july_02_35.php (accessed 18 March 2004).

—— (2003) 'Pacific Overtones and Bellicose Rhetoric', *Seapower Almanac 2003*. Online. Available HTTP: http://www.navyleague.org/sea_power/almanac_jan03_44.php (accessed 13 May 2004).

Gray, C.S. (1986) *Nuclear Strategy and National Style*, London: Hamilton Press.

—— (1990) *War, Peace and Victory: Strategy and Statecraft for the Next Century*, New York: Simon & Schuster.

—— (1992a) *House of Cards: Why Arms Control Must Fail*, New York: Cornell University Press.

—— (1992b) *The Leverage of Seapower: The Strategic Advantage of Navies in War*, New York: Free Press.

—— (1995) 'Puissance maritime, puissance continentale et la recherche de l'avantage stratégique', in H. Coutau-Bégarie (ed.) *La Lutte pour l'empire de la mer*, Paris: Economica.

—— (1999a) 'Inescapable Geography', in C. S. Gray and G. Sloan (eds) *Geopolitics, Geography and Strategy*, London: Frank Cass.

—— (1999b) *Modern Strategy*, Oxford: Oxford University Press.

Grousset, R. (1942; 1994; 3rd edn 2000) *Histoire de la Chine*, Paris: Payot.

Groves, D. G. and Hunt, L. M. (1980) *Ocean World Encyclopedia*, New York: McGraw-Hill.

Handel, M. I. (1992; 3rd edn 2001) *Masters of War: Classical Strategic Thought*, London: Frank Cass.

Hawkins, C.F. (2000) 'The Four Futures: Competing Schools of Military Thought inside the PLA'. Online. Available HTTP:http://taiwansecurity.org/IS/IS-0300-Hawkins.htm (accessed 10 September 2004).

He, D. (2003) 'The Last Campaign to Unify China: The CCP's Unrealized Plan to Liberate Taiwan, 1949–1950', in M. A. Ryan, D. M. Finkelstein and M. A. McDevitt (eds) *Chinese Warfighting*, New York: M. E. Sharpe.

Herrick, R. W. (1968) *Soviet Naval Strategy: Fifty Years of Theory and Practice*, Annapolis: Naval Institute Press.

Heuser, B. (2002) *Reading Clausewitz*, London: Pimlico.

Hezlet, A. (1967) *The Submarine and Sea Power*, New York: Stein and Day.

Holt, J. H.(1999) 'The China–Taiwan Military Balance', in W. L. Yang and D. A. Brown (eds) *Across the Taiwan Strait: Exchanges, Conflicts and Negotiations*, New York: Center for Asian Studies, St John's University, pp. 181–219. Online. Available HTTP: http://www.atimes.com/china/BA27Ad01.html (accessed 7 October 2004).

Hsü, I. C. Y. (1970; 5th edn 1995) *The Rise of Modern China*, New York: Oxford University Press.

Ikenberry, G. J. (2004) 'The End of the Neo-Conservative Moment', *Survival*, 46 (1): 7–22.

International Chamber of Commerce (ICC)/International Maritime Bureau (IMB) (2003) 'High Seas Terrorism Alert in Piracy Report', 29 January 2003. Online. Available HTTP: http://www.iccwbo.org/home/news_archives/2003/stories/piracy/20_report_2002.asp (accessed 7 October 2004).

International Herald Tribune (*IHT*) (2004) 'Air Power to Replace US Troops in the Pacific', 19 January.

International Institute for Strategic Studies (2000) 'Japan's Naval Power: Responding to New Challenges', *Strategic Comments*, 6 (8). Online. Available HTTP: http://www.iiss.org/newsite/stratcomsubarchive.php?scID=140 (accessed 18 March 2004).

—— (2003) *The Military Balance 2003–2004*, London: Oxford University Press.

Jervis, R. (1977) *Co-operation under the Security Dilemma*, ACIS Working Paper No. 4, Los Angeles: Center for Arms Control and International Security, University of California.

Johnston, A. I. (1995) *Cultural Realism: Strategic Culture and Grand Strategy in Chinese History*, Princeton: Princeton University Press.

Kagan, R. (2004) 'America's Crisis of Legitimacy', *Foreign Affairs*, 38 (2): 65–87.

Kahn, J. (2004) 'Hu Takes Control as Jiang Resigns', *International Herald Tribune*, 20 September 2004.

Kane, T. M. (2002) *Chinese Grand Strategy and Maritime Power*, London: Frank Cass.

Kang, D. C. (2003) 'Getting Asia Wrong: The Need for New Analytical Frameworks', *International Security*, 27 (4): 57–85.

Kanwa Intelligence Review Daily News (2003a) 'Deployment of SU30MKKs poses challenge to Taiwan's advantage', 15 May. Online. Available HTTP: http://www.kanwa.com/eindex.html (accessed 23 March 2004).

—— (2003b) 'Kilo 636 Project under Production', 4–12 August. Online. Available: HTTP: http://www.kanwa.com/eindex.html (accessed 7 October 2004).

—— (2003c) 'China Received More SU30MKK', 23–27 June 2003. Online. Available HTTP: http://www.kanwa.com/eindex.htm (accessed 13 May 2004).

Kaplan, B. (1999) 'China's Navy Today: Storm Clouds on the Horizon … or Paper Tiger?', *Seapower*, December 1999. Online. Available HTTP: http://www.navyleague.org/sea_power/chinas_navy_today.htm (accessed 13 May 2004).

Keay, J. (2000) *A History of India*, London: HarperCollins.

Keegan, J. (1993) *Battle at Sea: From Man-of-War to Submarine*, London: Pimlico.

Khalilzad, Z, Orletsky, D., Pollack, J., Pollpeter, K., Rabasa, A., Shlapak, D., Shulsky, A. and Tellis, A. (2001) *The United States and Asia: Towards a New US Strategy and Force Posture*, Santa Monica: RAND. Online. Available HTTP: http://www.rand.org/publications/MR/MR1315/ (accessed 20 May 2004).

Kondapalli, S. (2001) *China's Naval Power*, New Delhi: Knowledge World.

Lakshamanan, I. (2003) 'Diesel Engine May Have Sucked Out Submarine Oxygen', *Sydney Morning Herald*, 5 May 2003. Online. Available HTTP: http://www.smh.com.au/articles/2003/05/04/1051987611291.html (accessed 6 May 2003).

Landry, J. R. (2001) 'The Military Dimensions of Great Power Rivalry in the Asia-Pacific Region', in P. D. Taylor (ed.) *Asia and the Pacific: US Strategic Traditions and Regional Realities*, Newport, Rhode Island: Naval War College Press.

Lawrence, T. E. (1926; 5th edn 1976; 1988) *Seven Pillars of Wisdom*, London: Guild Publishing.

Leier, M. (2001) *World Atlas of the Oceans*, Buffalo, NY: Firefly Books.

Lewis, J. W. and Xue, L. (1988) *China Builds the Bomb*, Stanford: Stanford University Press.

—— (1994) *China's Strategic Seapower: The Politics of Force Modernisation in the Nuclear Era*, Stanford: Stanford University Press.

Lexington Institute (2004) *The Chinese Military: An Emerging Maritime Challenge*. Online. Available HTTP: http://www.lexingtoninstitute.org/defense/default.asp (accessed 1 September 2004).

Lim, R. (2003) *The Geopolitics of East Asia: The Search for Equilibrium*, London: Routledge.

Lindberg, M. and Todd, D. (2002) *Brown-, Green-, and Blue-Water Fleets: The Influence of Geography on Naval Warfare, 1861 to the Present*, Westport: Praeger.

Luttwak, E. (1987), *Strategy: The Logic of War and Peace*, Cambridge, Mass.: Harvard University Press.

Luttwak, E. and Koehl, S.L. (1991) *The Dictionary of Modern War*, New York: Grammercy Books.

Lynn-Jones, S. M. (1995) 'Offense–Defense Theory and its Critics', *Security Studies*, 4 (4): 660–91.

McDevitt, M. (2000) 'Ruminations about How Little We Know about the PLA Navy', paper delivered to the Conference on Chinese Military Affairs, National Defense University, Washington, DC, 10 October 2000. Online. Available HTTP: http://www.ndu.edu/inss/China-Center/paper14.htm (accessed 1 June 2004).

—— (2001) 'Roundtable Net Assessment – Objective Conditions versus the US Strategic Tradition', in P. D. Taylor (ed.) *Asia and the Pacific: US Strategic*

Traditions and Regional Realities, Newport, Rhode Island: Naval War College Press, pp. 101–6. Online. Available HTTP: http://www.nwc.navy.mil/apsg/nwcpapers.htm (accessed 12 October 2004).

McKittrick, M.S. (2003) *Submarines: Weapons of Choice for Future Conflict*, Arlington: Lexington Institute. Online. Available HTTP: http://www.lexingtoninstitute.org/navalstrike/default.asp (accessed 27 September 2004).

McPhedran, I. (2003) '"Dud" Subs Defeat US in Exercise', *Courier Mail*, 24 September.

Mahan, A. T. (1890; 1965) *The Influence of Sea Power upon History 1660–1783*, London: Methuen.

Mao Z. (1936; 1963) 'Problems of Strategy in China's Revolutionary War', in *Selected Military Writings of Mao Tse-Tung*, Peking: Foreign Languages Press.

—— (1938; 1963) 'Problems of Strategy in Guerrilla War against Japan' in *Selected Military Writings of Mao Tse-Tung*, Peking: Foreign Language Press.

Masson, P. (2002) *La Puissance maritime et navale au XXe siècle*, Paris: Perrin.

Mathews, W. (2003) 'Pentagon Wins Much, Not All, in Appropriations', *Defense News*, 13 October: 18.

Mearsheimer, J. J. (2001) *The Tragedy of Great Power Politics*, New York: Norton.

Menzies, G. (2002) *1421: The Year China Discovered the World*, London: Bantam Press.

Miller, D. (1998) *The Cold War: A Military History*, New York: St Martin's Press.

Miller, D. and Jordan, J. (1987) *Modern Submarine Warfare*, New York: Military Press.

Minnick, W. (2004) 'The Year to Fear for Taiwan: 2006', *Asia Times*, 10 April 2004. Online. Available HTTP: http://taiwansecurity.org/News/2004/AT-100404.htm (accessed 10 September 2004).

Moore, J. E. and Compton-Hall, R. (1987) *Submarine Warfare Today and Tomorrow*, Bethesda, Maryland: Adler & Adler.

Morgan, J. (1998) 'Anti-Submarine Warfare: A Phoenix for the Future', *Anti-Submarine Warfare*, 1 (1). Online. Available HTTP: http://www.fas.org/man/dod-101/sys/ship/docs/anti-sub.htm (accessed 2 February 2004).

Morgenthau, H. J. (1948; brief edn 1985; 6th edn 1993) *Politics among Nations: The Struggle for Power and Peace*, brief edition revised by K. W. Thompson, Boston: McGraw Hill.

Mulvenon, J. (2004) 'The PLA, Chen Shui-Bian, and the Referenda: The War Dogs that Didn't Bark', *China Leadership Monitor*, (10). Online. Available HTTP: http://www.chinaleadershipmonitor.org/20042/jm.html (accessed 30 September 2004).

Nathan, A. J. and Ross, R. S. (1997) *The Great Wall and the Empty Fortress: China's Search for Security*, New York: Norton.

Niquet, V. (1997) *Les Fondaments de la stratégie chinoise*, Paris: Economica.

—— (1999) *Sun Zi: L'Art de la guerre*, Paris: Economica.

Nitze, P. H. (1964; reprinted 1998) 'An Address', *Naval War College Review*, LI (1) (361) . Online. Available HTTP: http://www.nwc.navy.mil/press/Review/1998/winter/rtoc-w98.htm (accessed 24 September 2004).

Norris, R. S. and Kristensen, H. M. (2003) 'Chinese Nuclear Forces, 2003', *Bulletin of the Atomic Scientists*, 59 (6): 77–80.

O'Hanlon, M. (2000) 'Why China Cannot Conquer Taiwan', *International Security* 25 (2): 51–86.

Padfield, P. (1999; 2nd edn 2000) *Maritime Supremacy and the Opening of the Western Mind: Naval Campaigns that Shaped the Modern World, 1588–1782*, London: Pimlico.

Paul, T. V. (1994) *Asymmetric Conflicts: War Initiation by Weaker Powers*, Cambridge: Cambridge University Press.

People's Daily (2003) 'China to Cut PLA by 200,000 troops by 2005', 1 September 2003. Online. Available HTTP: http://taiwansecurity.org/News/2003/PD-090103.htm (accessed 17 June 2004).

Pillsbury, M. (2000) *China Debates the Future Security Environment*, Washington, DC: National Defense University Press. Online. Available HTTP: http://www.ndu.edu/inss/books/books%20-%202000/China%20Debates%20Future%20Sec%20Environ%20Jan%202000/pills2.htm (accessed 4 February 2004).

—— (2001) *China's Military Strategy Towards the US: A View from Open-Sources*, research paper prepared for the US–China Security Review Commission, 3 August 2001. Online. Available HTTP: http://www.uscc.gov/textonly/txrese.htm (accessed 7 October 2004).

Polmar, N. (2004a) 'Airborne ASW: A Critical Issue (Part 1)', *US Naval Institute Proceedings*, 130/4/1 (214): 88.

—— (2004b) 'Airborne ASW: A Critical Issue (Part 2)', *US Naval Institute Proceedings*, 130/5/1 (215): 88.

Pomfret, J. (2002) 'China to Buy 8 More Russian Submarines', *Washington Post*, 25 June. Online: Available HTTP: http://www.kanwa.com/eindex.html (accessed 9 October 2004).

Qiao L. and Wang X. (1999; French trans. 2003) *La Guerre sans limites*, translated by Hervé Denès, Paris: Rivages.

Raghuvanshi, V. (2003) 'US, Indian Navies Finish Major Exercises in the Arabian Sea', *Defense News*, 20 October: 15.

Ranft, B. and Till, G. (1983) *The Sea in Soviet Strategy*, Annapolis: Naval Institute Press.

Ratnam, G. (2004) 'US Navy, Marines Boost Ship Plans', *Defense News*, 22 February: 22.

Reid, M. (1998) *The US–Japan Security Relationship in the New Era*, Canberra: Strategic and Defence Studies Centre.

Rhem, K. T. (2004) 'US Realigning, Redeploying Military Forces in South Korea', American Forces Press Service, 2 September 2004. Online. Available HTTP: http://www.defenselink.mil/news/Sep2004/n09022004_2004090207.html (accessed 10 September 2004).

Rhodes, E. (1996) 'Sea Change: Interest-based vs Cultural-Cognitive Accounts of Strategic Choice in the 1890s', *Security Studies* 5 (4): 73–124.

Rice, C. (2000) 'Promoting the National Interest', *Foreign Affairs,* 79 (1): 45–62.

Richardot, P. (2002) *Les États-Unis hyperpuissance militaire*, Paris: Economica.

Rosinski, H. (1939) 'Command of the Sea', in B. M. Simpson III (ed.) (1977) *The Development of Naval Thought: Essays by Herbert Rosinski*, Newport: Naval War College Press.

Ross, R. S. (1999) 'The Geography of Peace: East Asia in the Twenty-First Century', *International Security*, 23 (4): 81–118; reprinted in M. E. Brown, O. R. Coté, Jr, S. M. Lynn-Jones and S. E. Miller (eds) (2000) *The Rise of China*, Cambridge, Mass.: MIT Press, pp. 167–204.

—— (2000) 'The 1995–96 Taiwan Strait Confrontation: Coercion, Credibility, and the Use of Force', *International Security*, 25 (2): 87–123.

REFERENCES

—— (2002) 'Navigating the Taiwan Strait: Deterrence, Escalation Dominance, and U.S.–China Relations', *International Security*, 27 (2): 48–85.

Roy, D. (2003) 'China's Reaction to American Predominance', *Survival*, 45 (3): 57–78.

Ryan, M. A., Finkelstein, D. M. and McDevitt, M. A. (2003) *Chinese Warfighting: The PLA Experience since 1949*, New York: M. E. Sharpe.

Sawyer, R. D. (1993) *The Seven Military Classics of Ancient China*, Boulder: West-view Press.

Schelling, T. C. (1966) *Arms and Influence*, New Haven: Yale University Press.

Scobell, A. (2003) *China's Use of Military Force: Beyond the Great Wall and the Long March*, Cambridge: Cambridge University Press.

Seapower (2003) 'Sub Director Foresees "Revolutionary" Power of SSGNs', *Seapower*, July. Online. Available HTTP: http://www.navyleague.org/sea_power/jul_03_issue.php (accessed 2 February 2004).

Shambaugh, D. (2000) 'Sino-American Strategic Relations: From Partners to Competitors', *Survival*, 42 (1): 97–115.

—— (2002) *Modernizing China's Military: Progress, Problems and Prospects*, Berkeley: University of California Press.

—— (2003) 'China's New High Command', in S. J. Flanagan and M. E. Marti (eds) *The People's Liberation Army and China in Transition*, Washington, DC: National Defense University Press. Online. Available HTTP: http://www.ndu.edu/inss/books/Book_titles.htm (accessed 8 October 2004).

Shen Z., Zhang H. and Zhou X. (1998) '21st Century Naval Warfare', in M. Pillsbury (ed.) *Chinese Views of Future Warfare*, Washington, DC: NDU Press. Online. Available HTTP: http://www.ndu.edu/inss/books/Book_titles.htm (accessed 7 October 2004).

Sherman, J. (2003) 'US to Create Command, Boost Anti-Sub Spending', *Defense News*, 22 September: 6.

Shlapak, D. A., Orletsky, D. T. and Wilson, B. (1999) *Dire Strait? Military Aspects of the China–Taiwan Confrontation and Options for US Policy*, Santa Monica: RAND. Online. Available HTTP: http://www.rand.org/publications/MR/MR1217/ (accessed 7 April 2004).

Spector, R. H. (1985) *Eagle against the Sun: The American War with Japan*, New York: Free Press.

Stefanick, T. (1987) *Strategic Antisubmarine Warfare*, Lexington: Lexington Books.

Stokes, M. A. (1999) *China's Strategic Modernization: Implications for the United States*, Carlisle, Pennsylvania: Strategic Studies Institute, US Army War College. Online. Available HTTP: http://www.carlisle.army.mil/ssi/pubs/pubResult.cfm/hurl/PubID=74/ (accessed 23 September 2004).

—— (2000) 'China's Military Space and Conventional Theater Missile Development: Implications for the Security of the Taiwan Strait', in S. M. Puska (ed.) *People's Liberation Army after Next*, Carlisle, Pennsylvania: Strategic Studies Institute.

Storey, I. and You J. (2004) 'China's Aircraft Carrier Ambitions: Seeking Truth from Rumors', *Naval War College Review*, LVII (1): 77–93.

Swaine, M. D. (1992) *The Military and Political Succession in China: Leadership, Institutions and Beliefs*, Santa Monica: RAND.

Swaine, M. D. and Tellis, A. (2000) *Interpreting China's Grand Strategy: Past, Present and Future*, Santa Monica: RAND.

Swanson, B. (1982) *Eighth Voyage of the Dragon: A History of China's Quest for Seapower*, Annapolis: Naval Institute Press.

Taylor, A. J. P. (1954; 1971) *The Struggle for Mastery in Europe*, Oxford: Oxford University Press.

Tellis, A. J., Chung, M. L., Mulvenon, J., Purrington, C. and Swaine, M. D. (1998) 'Sources of Conflict in Asia', in Z. Khalilzad and I. O. Lesser (eds) *Sources of Conflict in the 21st Century: Regional Futures and US Strategy*, Santa Monica: RAND.

Teo, L. (2004) 'Asean Doesn't Want Taipei to Destabilize Region', *Straits Times*, 22 June. Online. Available HTTP: http://taiwansecurity.org/ST/2004/ST-220604–2.htm (accessed 27 September 2004).

Terrill, R. (2003) *The New Chinese Empire*, Sydney: UNSW Books.

Thompson, L. B. (2000) 'A Decade of Decision Looms for the Undersea Fleet', Issue Brief, Lexington Institute, 12 September. Online. Available HTTP: http://www.lexingtoninstitute.org/defense/underseaflt.htm (accessed 23 May 2004).

—— (2001) *Aircraft Carrier (In)vulnerability*, Arlington: Lexington Institute Naval Strike Forum. Online. Available HTTP: http://www.lexingtoninstitute.org/naval-strike/default.asp (accessed 28 September 2004).

Tien, C. (1992) *Chinese Military Theory, Ancient and Modern*, Oakville, Ontario: Mosaic Press.

Tomkins, J. (2003) 'How US Strategic Policy is Changing China's Nuclear Plans', *Arms Control Today*, January–February. Online. Available HTTP: http://www.armscontrol.org/act/2003_01–02/tompkins_janfeb03.asp (accessed 8 October 2004).

Tripplet, W. C. II (2000) 'Potential Applications of PLA Information Warfare Capabilities to Critical Infrastructures', in S. D. Puska (ed.) *People's Liberation Army after Next*, Washington, DC: Strategic Studies Institute.

Tsang, I. (2003) 'China's Navy Floats a Warning to Taiwan', *Asia Times*, 25 November 2003. Online. Available HTTP: http://taiwansecurity.org/News/2003/AT-251103.htm (accessed 11 May 2004).

United States Department of Defense (USDoD) (1999) 'Report to Congress on the Security Situation in the Taiwan Strait,' 26 February 1999. Online. Available HTTP: http://www.fas.org/news/taiwan/1999/twstrait_02261999.htm (accessed 4 October 2004).

—— (2000) 'Joint Vision 2020: America's Military: Preparing for Tomorrow', Chairman of the Joint Chiefs of Staff (CJCS), Washington, DC. Online. Available HTTP: http://www.dtic.mil/jointvision/jvpub.2.htm (accessed 14 May 2004).

—— (2001) 'Quadrennial Defense Review Report', Online. Available HTTP: http://www.defenselink.mil/pubs/archive.html (accessed 2 June 2004).

—— (2002) 'Annual Report on the Military Power of the People's Republic of China', Washington, DC. Online. Available HTTP: http://www.defenselink.mil/news/Jul2002/d20020712china.pdf (accessed 2 May 2004).

—— (2003) 'Annual Report on the Military Power of the People's Republic of China', Washington, DC. Online. Available HTTP: http://www.defenselink.mil/pubs/200307730chinaex-1.pdf (accessed 2 April 2004).

—— (2004a) 'Prepared Testimony of the Secretary of Defense Donald H. Rumsfeld, for the Senate and House Armed Services Committees, Wednesday February 4,

2004'. Online. Available HTTP: http://www.defenselink.mil/speeches/2004/sp20040204-secdef0842.html (accessed 17 September 2004).

—— (2004b) 'Annual Report on the Military Power of the People's Republic of China', Washington, DC. Online. Available HTTP: www.defenselink.mil/pubs/d20040528PRC.pdf (accessed 5 October 2004).

United States General Accounting Office (2001a) 'Force Structure: Options for Enhancing the Navy's Attack Submarine Force', Report to the Subcommittee on Seapower, Committee on Armed Services, US Senate, November. Online. Available HTTP: http://www.gao.gov/new.items/d0297.pdf (accessed 18 March 2004).

—— (2001b) 'Navy Acquisitions : Improved Littoral War-fighting Capabilities Needed', Report to the Chairman and Ranking Minority Member, Subcommittee on Military Research and Development, Committee on Armed Services, House of Representatives, May 2001. Online. Available HTTP: http://www.gao.gov/new.items/d01493.pdf (accessed 18 March 2004).

—— (2004) 'Space-Based Radar Effort Needs Additional Knowledge before Starting Development', Report on Defense Acquisitions to Congressional Committees No. GAO-04-759, July 2004. Online. Available HTTP: http://www.gao.gov/new.items/d04759.pdf (accessed 17 September 2004).

United States Naval Doctrine Command (1998) 'Littoral Anti-Submarine Warfare Concept'. Online. Available HTTP: http://www.fas.org/man/dod-101/sys/ship/docs/aswcncpt.htm#2 (accessed 24 September 2004).

United States Navy (2002) 'Vision ... Presence ... Power: A Programme Guide to the US Navy, 2002 Edition', Washington, DC. Online. Available HTTP: http://www.chinfo.navy.mil/navpalib/policy/vision/vis02/vpp02-ch1c.html (accessed 14 May 2004).

—— (2004) 'Carrier Strike Group', Fact File. Online. Available HTTP: http://www.chinfo.navy.mil/navpalib/ships/carriers/powerhouse/cvbg.html (accessed 30 September 2004).

—— (undated) 'SSGN: A Transformational Force for the US Navy'. Online. Available HTTP: http://www.chinfo.navy.mil/navpalib/cno/n87/usw/issue_13/ssgn.htm (accessed 27 September 2004).

United States Senate (1999) 'Department of Energy, FBI, and Department of Justice Handling of the Espionage Investigation into the Compromise of Design Information on the W-88 Warhead, Statement by Senate Governmental Affairs Committee Chairman Fred Thompson (R-TN), Senate Governmental Affairs Committee Ranking Minority Member Joseph Lieberman, (D-CT) August 5, 1999'. Online. Available: http://www.senate.gov/~gov_affairs/080599_china_espionage_statement.htm (accessed 6 October 2004).

—— (2004) 'Statement by Richard P. Lawless, Deputy Undersecretary of Defense', US Senate Foreign Relations Committee Subcommittee on East Asian and Pacific Affairs Hearing on US–China Relations: Status of Reforms in China, 22 April 2004. Online. Available HTTP: http://foreign.senate.gov/hearings/2004/hrg040422p.html (accessed 9 September 2004).

Van der Vat, D. (1995) *Stealth and Sea: The History of the Submarine*, New York: Houghton Mifflin.

Van Evera, S. (1999) *Causes of War: Power and the Roots of Conflict*, New York: Cornell University Press.

Vego, M. N. (1999) *Naval Strategy and Operations in Narrow Seas*, London: Frank Cass.

Waltz, K. N. (2003) 'More May Be Better', in S. D. Sagan and K. N. Waltz, *The Spread of Nuclear Weapons: A Debate Renewed*, New York: Norton.

Weeks, S. B. and Meconis, C. A. (1999) *The Armed Forces of the USA in the Asia-Pacific Region*, Sydney: Allen & Unwin.

Weir, G. E and Boyne, W. J. (2003) *Rising Tide: The Untold Story of Russian Submarines that Fought the Cold War*, New York: Basic Books.

Whitson, W. W. and Chen-hsia, H. (1973) *The Chinese High Command: A History of Communist Military Politics, 1927–71*, New York: Praeger.

Wolf, J. Jim (2003) 'China Continues Arms Shopping Spree', *The Age*, 27 September 2003. Online HTTP: http://www.theage.com.au/text/articles/2003/09/26/1064083188091.htm (accessed 28 September 2003).

Wolfe, T. W. (1973) 'Soviet Naval Interaction with the United States and its Influence on Soviet Naval Developments', in M. MccGwire (ed.) *Soviet Naval Developments: Capability and Context*, New York: Praeger.

Woodrow Wilson International Center (1958) *First Conversation of N. S. Khrushchev with Mao Zedong Hall of Huaizhentan [Beijing]*, Cold War International History Project Virtual Archive, Document 1. Online. Available HTTP: http://wwics.si.edu/index.cfm?fuseaction=library.document&topic_id=1409&id=14658 (accessed 11 October 2004).

Wortzel, L. M. (2003) 'The Beiping–Tianjin Campaign of 1948–49: The Strategic and Operational Thinking of the People's Liberation Army', in M. A. Ryan, D. M. Finkelstein, and M. A. McDevitt (eds) *Chinese Warfighting: The PLA Experience since 1949*, New York: M. E. Sharpe.

Wright, D. C. (2002) 'The Northern Frontier', in D. A. Graff and R. Higham (eds) *A Military History of China*, Boulder: Westview Press.

Wu, D. (2004) 'Russia Could Make Subs for Taipei', *Taipei Times*, 26 June 2004. Online. Available HTTP: http://taiwansecurity.org/TT/2004/TT-260604.htm (accessed 7 September 2004).

Wu, T. (2003) 'Taiwan Eyes Submarines, Anti-missile System', Reuters, 30 August 2003. Online. Available HTTP: www.taiwansecurity.org/Reu/2003/Reuters-083003.htm (accessed 5 June 2004).

Wylie, J. C. (1967) *Military Strategy: A General Theory of Power Control*, New Brunswick: Rutgers University Press.

You, J. (1999) *The Armed Forces of China*, Sydney: Allen & Unwin.

INDEX

Academy of Military Sciences 27–8, 133
Advanced Deployable System (ADS) 37
Aegis system 65–7, 136, 138
Afghanistan 61, 63, 96, 97, 111, 117, 118
aircraft carriers 3, 11, 20, 45, 54, 57, 59,
    60, 61, 66, 63, 127–8, 159, 171;
    Chinese 45, 67; as PLA targets 20, 77,
    98–100, 102, 103, 104, 107, 108, 128,
    153, 156, 170, 171; as Soviet targets
    45, 79, 117, 128, 129; US 2, 45, 54,
    57, 58, 59, 60, 61, 63, 79, 99, 102, 104,
    117–18, 159; vulnerability of 45,
    64–7, 85, 90–1, 99, 100, 101
Alaska 36, 57
Aleutian Islands 41
Allison, Graham, and Zelikow, Philip
    11–12, 13, 115
Andersen Air Force Base 59
anti-submarine warfare (ASW) 3, 16,
    36–8, 39, 50, 51, 52, 63, 76, 81, 89, 92,
    93, 94, 100, 103, 138, 139, 169, 171;
    US capabilities 108–12; *see also*
    submarine warfare
Apra Harbour 59
Argentina 154
ASEAN 165
asymmetric warfare 10, 64, 66, 80, 98,
    101, 134–7, 148, 153–66, 173, 174
Atsugi 58
Australia 16, 39, 100

Bashi Channel 146
Bay of Bengal 39
Beaufre, André 73, 96, 98
blockade: historical instances of 50, 51,
    52, 73, 170–1; against Taiwan 20, 48,
    49–51, 52–3, 55, 56, 140–2, 164;
    theory of 69, 86, 22
Bonin Islands 42

Booth, Ken 13, 113–15, 133
Borik, Frank 137–8
Børresen, Jacob 88–9, 90
Brodie, Bernard 9, 22, 64, 71, 72, 74, 79,
    80, 82, 91–2, 103, 158, 167–8
Bull, Hedley 3–4, 25
Burles, Mark, and Shulsky, Abram
    161–2, 163
Bush, George W. 30
Bushnell, David 71

Cambodia 162
Camp Butler 58
Camp Zama 59
CAPTOR 37, 38
Carolina Islands 42
Carr, E.H. 2, 175–6
Castex, Raoul 23, 72, 73, 74–5, 76, 77,
    78, 79–82, 83, 84, 124, 129, 136, 146,
    148, 167, 171, 173
Central Military Commission 13, 27, 41,
    47–8, 53, 117
Chen Shui-bian 164, 176, 177
Chiang Kai-Shek (Jiang Jieshi) 51
China: ballistic missiles 10, 31, 32, 34,
    48, 49, 63, 66, 106, 142, 146, 152, 155,
    172; continental strategy 7, 20, 21, 24,
    42, 43, 121, 134, 143–6; and East
    Asia 21, 40, 42; and Japan 6, 25, 29,
    62, 89, 169, 176; maritime and naval
    strategy 4, 5, 6, 20, 25, 26, 40, 41–56,
    69, 73, 74, 76, 77, 86–7, 96–8, 99, 100,
    102, 113, 117, 118, 122–3, 126–7, 129,
    130, 131, 132, 141, 146–7, 155–6, 158,
    165, 172; nuclear deterrence 5, 30–2,
    33–6, 38, 40, 41, 54, 154, 155, 174–5;
    and Russia 120; sovereignty claims
    4–5, 28–9, 41, 44, 45, 162; and Soviet
    Union 23, 24, 161, 162, 163, 169;

LaVergne, TN USA
11 November 2010
204555LV00002B/15/P